Palgrave Studies in Institutions, Economics and Law

Law and Economics is an interdisciplinary field of research that has emerged in recent decades, with research output increasing dramatically and academic programmes in law and economics multiplying. Increasingly, legal cases have an economic dimension and economic matters depend on rules and regulations. Increasingly, economists have realized that "institutions matter" because they influence economic activities. Increasingly, too, economics is used to improve our understanding of how institutions and how legal systems work. This new Palgrave Pivot series studies the intersection between law and economics, and addresses the need for greater interaction between the two disciplines.

More information about this series at
http://www.palgrave.com/gp/series/15241

Antonio Manganelli • Antonio Nicita

The Governance of Telecom Markets

Economics, Law and Institutions in Europe

palgrave
macmillan

Antonio Manganelli
European University of Rome
Rome, Italy

Antonio Nicita
Libera Università Maria SS. Assunta
Rome, Italy

ISSN 2662-6535 ISSN 2662-6543 (electronic)
Palgrave Studies in Institutions, Economics and Law
ISBN 978-3-030-58159-6 ISBN 978-3-030-58160-2 (eBook)
https://doi.org/10.1007/978-3-030-58160-2

This Palgrave Macmillan imprint is published by the registered company Springer Nature Switzerland AG.
The registered company address is: Gewerbestrasse 11, 6330 Cham, Switzerland

Preface

The governance of telecom markets results from a complex dynamic interaction between technology, economics and the law. Disruptive innovations, swift market dynamics and policy shifts exacerbate this complexity, making it increasingly challenging to get a sound understanding of the sector, its many dimensions and coevolutionary patterns.

Our intent is to provide, in one succinct book, a critical comprehensive summary of the coevolution of telecom markets, rules and public institutions over the last 25 years, with a focus on the challenges that regulators and policy makers have been facing. Even if the perspective of the book is European (as we examine EU law), most of the economic and institutional issues addressed are common to all telecom markets in advanced economies.

The book's introduction deals with *the Long Wave of Telecom Markets Liberalisation*. It provides a conceptual and thematic framework to the following analysis but could be also read as a stand-alone paper. The ensuing chapters expand on specific arguments raised in the introduction and are grouped in two parts: I. Pro-competitive and Pro-Consumer Rules, and II. Promotion of Investments and Technological Innovation. In addition, thematic boxes are included throughout the book to elucidate some specific relevant issues.

The book addresses some traditional fundamental topics in the telecom regulation literature, as well as some hot-button topics in the current policy debate, e.g., development of 5G and ultrafast broadband networks, the relationship between investments and competition, the digitalisation of the sector and the OTT's role. We only briefly touch upon the wider digital economy and the challenges posed by digital platforms, as this will be the subject of a specific forthcoming publication dealing with *a regulatory approach to digital capitalism*.

This book is based on our academic research as well as on our professional experience as civil servants in regulatory and competition authorities at national and EU level. For many fruitful discussions around these topics, held in different occasions over the years, we would like to thank:

Giuliano Amato, Laura Ammannati, Filippo Arena, Filippo Belloc, Delphine Bernet-Travert, Erik Bohlin, Oscar Borgogno, Marco Botta, Marc Bourreau, Eric Brousseau, Chiara Caccinelli, Carlo Cambini, Martin Cave, Giovanni Cazora, Giuseppe Colangelo, Giorgio Corda, Alberta Corona, Andrea Coscelli, Stefano da Empoli, Massimo D'Antoni, Maurizio Decina, Marco Delmastro, Alexandre De Streel, Antonio De Tommaso, Fabiana di Porto, Annalisa D'Orazio, Elisabeth Dornetshumer, Valeria Falce, Davide Gallino, Michael Grenfell, Annegret Groebel, Paula Gori, Stefano Gori, Steffan Hoernig, Martin Holterman, Sharon Horwitz, Jorge Infante, Panos Karaminas, Laszlo Kasa, Celine Kauffmann, Gavin Knott, James Lambert, Francesco Lo Passo, Paolo Lupi, Alain Marciano, Mauro Martino, Sandro Mendonça, Emanuela Michetti, Aldo Milan, Giorgio Monti, Giulio Napolitano, Pier Luigi Parcu, Antonio Perrucci, Anna Pietikainen, Anna Renata Pisarkiewicz, Andrea Pezzoli, Giovanni Pitruzzella, Matej Podbevsek, Gerard Pogorel, Steve Preece, Augusto Preta, Lorenzo Principali, Giovanni Ramello, Nicoletta Rangone, Maria Alessandra Rossi, Antonio Sassano, Giovanni Santella, Marco Scialdone, Vincenzo Valentini, Roberto Viola, Dirk Walpuski, Anthony Whelan.

Rome, Italy — Antonio Manganelli
Rome, Italy — Antonio Nicita

Contents

About the Authors

Antonio Manganelli, Ph.D. is Adjunct Professor of Regulation and Innovation Policy at the European University of Rome (Italy) and Competition Law and Policy at the University of Siena (Italy). He also works as managing director of the DEEP-IN research network. Manganelli has served in different public institutions, i.e., the Italian Regulatory Authority for Telecom, Media, and Postal sectors (AGCOM), the UK Competition and Markets Authority (CMA), the Office of the Body of European Regulators for Electronic Communications (BEREC). He has previously held academic positions at the European University Institute (EU) and holds a Ph.D. in Law and Economics from the University of Siena.

Antonio Nicita, Ph.D. is Full Professor of Economic Policy at LUMSA University (Italy) and visiting scholar at the Center for Research in Regulated Industry at the Rutgers Business School, Usa. He has served as Commissioner of the Italian Regulatory Authority for Telecom, Media, and Postal sectors (AGCOM) since 2014. Nicita is also member of the OECD Steering Group in Regulation & Emerging Technologiesat the Organisation for Economic Development and Cooperation (OECD). Nicita was Visiting Fulbright Professor at Yale University (USA), Visiting Scholar at the University of Paris X-Nanterre (France) and at the European University Institute (EU), and Visiting Researcher at the University of Cambridge (UK). He holds an M.Sc. in Economic and Social Sciences from the Bocconi University in Milan, and a Ph.D. in Economics from the University of Siena.

List of Figures

List of Tables

List of Boxes

CHAPTER 1

Introduction: The Long Wave of Telecom Market Liberalisation

Abstract For many years, natural monopoly and universal service arguments were used as key theoretical reasons for maintaining public legal monopolies in national telecom markets. At the end of the'80s, legal barriers started to be perceived as inappropriate and economically unsustainable and, in the EU, telecom liberalisation was introduced in accordance with existing treaties' objectives, i.e., market integration and competition enhancement. The removal of exclusive rights was a necessary but not sufficient condition to ensure an effective entry of new competitors. Economic regulation was further needed to address the competitive advantages enjoyed by former monopolies. Pro-competitive market outcomes, however, have varied considerably from country to country in Europe, despite a harmonised framework. This is due to complex evolutionary relationships among different liberalisation "policy variables". Moreover, a crucial policy debate has revolved around the interdependence of liberalisation and privatisation, and whether or not privatisation of incumbent firms could have an additional pro-competitive impact.

Keywords Telecommunications economics • Monopoly • Liberalisation • Privatisation • Competition

A. Manganelli, A. Nicita, *The Governance of Telecom Markets*,
Palgrave Studies in Institutions, Economics and Law,
https://doi.org/10.1007/978-3-030-58160-2_1

1.1 Telecommunications Economics and the Policy Case for Legal Monopolies

The telecom sector is defined as a *network industry*, i.e., an industry characterised by a collection of connected nodes, on both supply and demand sides. Services or goods are "delivered" among nodes to end-users: delivery connections are one-way directional in some network industry markets (e.g., natural gas, water, cable TV, broadcasting), while in others, such as phone and internet, connections are two-way directional.

A crucial economic feature of network industries, and telecom markets in particular, is *network externalities,* sometimes defined as *demand-sided scale economies,* due to which: (a) the overall value of a given network grows as users of the network increase, and (b) the marginal value that network users receive from their subscription grows as the number of reachable users increases.[1]

Networks could be either physical or virtual. Telecom networks, delivering services to consumers, are pervasive physical infrastructures.[2] Building and managing a telecom network infrastructure require high fixed costs (i.e., costs not dependent on the level of service provision), which underly *economies of scale*, that is, a continuous decline of the long-run average costs of production as the quantity produced increases. Moreover, telecom production cost is also influenced by *economy of density* for the provision of services, where the average cost of production decreases as the density of customers increases. All these elements in turn imply that a single firm supplying the whole market demand has lower production costs compared to the overall costs that two or more producers would have.

At the early stages of the industry, this economic framework originally led to the *natural monopoly argument*, where, for purely technological reasons, a monopoly is more cost-efficient than multi-firm production. In a single-product industry, scale economies are a sufficient (but not necessary) condition to have a natural monopoly, whereas in a multi-product industry, such as telecommunications, the sufficient condition for a natural monopoly is given by the concept of sub-additivity. Applied to the cost function, this means that, regardless of which combination of desired outputs is chosen, it is cheaper for a single firm to produce that combination.[3] This also implies the existence of *economies of scope*, another feature of the

[1] Shy (2001).

[2] Spulber and Yoo (2009).

[3] Baumol (1977).

telecom industry and those sectors with multi-purpose networks, where multiple services can be provided simultaneously. Economies of scope entail the average total cost of production to decrease as a result of increasing the number of different services produced.[4]

Hence, the traditional thought was that "the presence of natural monopoly characteristics often means that competition cannot be relied upon to provide the socially optimal outcome, and some form of government intervention in these industries may be desirable."[5] In the next section, we will see how technological innovation led to a paradigmatic change regarding the natural monopoly argument in telecom markets, favouring the idea of multiple competing network operators. However, traditionally, the search for a solution to the economic problem of natural monopoly and to the political issue related to a pervasive provision of telecom services at affordable prices (*universality of service*) led policymakers to provide one single company with exclusive and special rights, i.e., a legal monopoly. The only company allowed to lawfully operate in the market was required to guarantee services to the entire population at fair and equitable prices. The legal monopoly situation took two alternative forms, not dependent on economic rationale, but rather on regulatory legacy and political reasons: (a) state-owned national monopolies, as occurred in almost all EU and OECD countries; (b) private (local) monopolies, subject to public regulation, as occurred in the US.

Regardless of the ownership and control of companies, all legal monopolists were responsible for both building and maintaining the network and for delivering the final service to customers, meaning that "natural monopolies in networks supported monopolies in services."[6] This value chain setting is defined as *vertical integration* and is further explained by another economic characteristic of the telecom sector: different market segments of telecom markets are characterised by complementarities, making vertical integration cost-efficient (and vertical separation subject to opportunity costs). However, end-to-end vertical integration was used by policy-makers also to extend the scope of reserved activities, in order to include additional profitable services and products, e.g., the legal monopolies on phone terminals and phone directories, which was void of any economic rationale. Moreover, end-to-end integration allowed complete control of the network and services by the state for reasons of national security (national interest argument).

[4]Sharkey (1982).
[5]OECD (2000).
[6]Hellwig (2009).

Due to the economics of the sector and other political arguments, for many years the European telecom markets were indeed legally protected national monopolies, where only one vertically integrated company in each country could lawfully operate. Monopolistic market structures were also chosen based on the consideration that a free market would not have guaranteed the socially desired *universality of service*: it was (and still is) considered essential to provide basic communication services at an affordable price to all. In legal terms, communication services satisfy fundamental needs of customers and citizens, in order to pursue the constitutional principle of social inclusion. In economic terms, a universal service policy maximises the positive direct and indirect impact of telecom services on economic growth and development, and internalise their extensive positive externalities to the overall economy and society. As a matter of fact, many empirical analyses recognise a positive causal effect of telecom networks and services diffusion on GDP growth.[7] This effect increases more than proportionally as networks and services penetration increase and become significant once a *critical mass* is reached. Critical mass appears sometime to be near universal service level.[8]

However, profitability in the telecom industry is differentiated according to the demographic (and geographic) characteristics of each area, also because of economies of density. There are therefore more profitable, less profitable and entirely unprofitable areas. Under a legal monopoly regime, the provision of service in unprofitable areas is financed by cross subsidies. Under a cross subsidies regime, monopoly gains in profitable areas, which are significantly higher than what they would be in a competitive market, compensate losses in unprofitable areas. Whereas, in a free or liberalised market, new operators would enter solely on profitable areas and/or provide solely profitable services. This selective entry or *cream-skimming* strategy typically generates two main consequences: (a) development of competition in the profitable market segments, while a monopoly market structure is maintained in all other areas; and (b) a reduction of the expected gains of the incumbent (skimmed by new entrants), preventing cross-subsidisation of services in loss-making areas. This, in turn, would imply a market reaction from the incumbent in unprofitable areas: it would either (i) exit (or reduce services) or (ii) increase prices compared to the previous affordable levels. Both outcomes are considered undesirable

[7] Czernich et al. (2011).

[8] Roller and Waverman (2001).

under a universal service approach and thus reinforced the case for a monopolistic market structure.

In most OECD countries, natural monopoly, national interest and the universal service arguments were used as the key theoretical reasons for adopting and maintaining public legal monopolies in national telecom markets. Both the social importance and the economic relevance of the sector in terms of GDP and public employment (political leverages) led most national European policy-makers to prefer direct public control over those monopolies. [9]

1.2 The Liberalisation of Telecom Markets: Economic Rationale and Legal Tools

When we talk today about telecom monopolies, we probably think of the old times of analogue telephones, public telephone kiosks or even telegraphs. Indeed, telecom industry has been subject to a continuous evolution, both under economic and technological perspectives.

Technological innovation has accelerated exponentially over the past few decades. The development of wireless communications and the internet has completely reshaped service provision, whereas digitalisation has deeply transformed the entire sector, extending its scope, and setting aside the rigid correspondence between types of networks and types of services. Today, one digitised "general purpose" *electronic communication* transmission network can convey a wide range of services, which in the analogue world either belonged to different sectors (e.g., telecom, audio-visual media, and information technology) or simply did not exist. This process has been (partially) acknowledged by European law which has been moving toward a single regulatory framework for all *electronic communications* infrastructure and associated services.[10]

If *digitalisation, convergence* and *mobility* have been the key technological concepts of the telecom industry's momentous evolution over the last three decades, from an economic policy perspective we have witnessed a likewise momentous transformation, leading to the shift *from monopoly to full competition.*[11] From the end of the '80s, and more extensively from the '90s, legal barriers protecting monopolies progressively started to be

[9] Nicita and Belloc (2016).

[10] Starting from the 1999 review. See European Commission (1999).

[11] Recital 1 Directive 2002/21/EC on a common regulatory framework for electronic communications networks and services (Framework Directive—FD).

perceived as politically undesirable and economically unsustainable. Without a doubt, liberalisation was strongly supported and nurtured by the liberal political wave animating the US and UK policy agenda at that time. Furthermore, the evolution of economic theory and technological evolution gave a crucial contribution.

On the one hand, the natural monopoly argument began losing its strength as technological innovation and digitalisation had dramatically decreased telecommunication network deployment costs and increased telecom service cost-efficiency. On the other hand, the economic and policy debates focused more and more on the inefficiencies of monopolistic market structures, in particular:

- dynamic inefficiencies, since a monopolistic firm has lower incentives to innovate compared to a company in a competitive context;[12]
- *x-inefficiency*, according to which a monopoly has higher average costs due to lack of competitive pressure;[13]
- allocative inefficiencies, i.e. the social cost of monopolies due to:

  - the so-called *deadweight loss*, that is the loss sustained by consumers not buying the product/service because of its higher price (monopoly price, compared to a competitive price), which is not offset by gains to the monopolist, and
  - the opportunity costs of those resources directed by monopolists to efforts to obtain or keep monopoly profits.[14]

In the European Union, liberalisation policy in telecom markets was introduced by the European Commission at the end of the '80s with a Green Paper on the development of the common market for telecommunications services and equipment [15] and implemented in accordance with the objectives of market integration and competition, enshrined in the treaties of the

[12] von Hayek (1978).

[13] Leibenstein (1966).

[14] Posner (1975).

[15] European Commission (1987). The 1987 Green Paper on the development of the common market for telecommunications services and equipment had 3 main objectives: (a) promote open and competitive markets, (b) build an EU internal market, and (c) pursue EU citizen interest.

European Community.[16] EU telecom market liberalisation was then based on existing legal provisions, i.e., on art. 106 TFEU, according to which EU Member States must refrain from adopting measures contrary to the EU competition rules, neither for public undertakings nor for undertakings benefiting from special and exclusive rights.[17] Moreover, the Commission is enabled to adopt decisions or directives to enforce that provision, without introducing any additional obligation for the Member States.[18] In doing this, the Commission has an exclusive enforcement power, with no duty to include EU co-legislators (EU Parliament and EU Council).

Art. 106 has actually been in force since 1957,[19] therefore, the liberalisation legislative base was not a specific new act, but an overhaul in the interpretation and implementation of existing principles.[20] Indeed, the first group of liberalisation directives were adopted by the European Commission in opposition to the economic policy followed at that time by most of the EU Member States. These directives clarified the new EU policy, considering national legislations granting special and exclusive rights to a single telecom company as contrary to the *competition principle* enshrined in the EU treaties. The competition principle, therefore, served as an external constraint to anti-competitive (intrusive) state intervention.[21] This approach resulted in a group of directives which progressively required Member States to break the national monopolistic legal frameworks in telecom markets, thus allowing other companies to enter the market. This milestone was placed in 1996 with the *Full Competition Directive* [22] which completely liberalised the telecom industry, reaching the core service of the industry: the fixed telephony market.

[16]The European Community and Union treaties have always been based on the liberal principles of competition, control of state aid, and the free movement of people, goods and services. Specifically, the competition rules applicable to companies are based on articles 101 and 102 TFEU, which can also be applied to public enterprises and companies with special or exclusive rights, art. 106 (2) TFEU.

[17]Art. 106(1) TFEU.

[18]Art. 106(3) TFEU.

[19]The Treaty establishing the European Economic Community (TEC) was signed in Rome on 25 March 1957.

[20]Napolitano (2005).

[21]Amato and Laudati (2001).

[22]Directive 96/19 (full competition directive) prohibited exclusive rights in fixed phone services, which was the core of the telecom industry. The previous directives addressed other telecom markets, such as the Terminals Directive (88/301/EEC), Service Directive (90/388), Satellite Communications Directive (94/46), Mobile Services Directive (96/2) and Cable Services Directive (95/51).

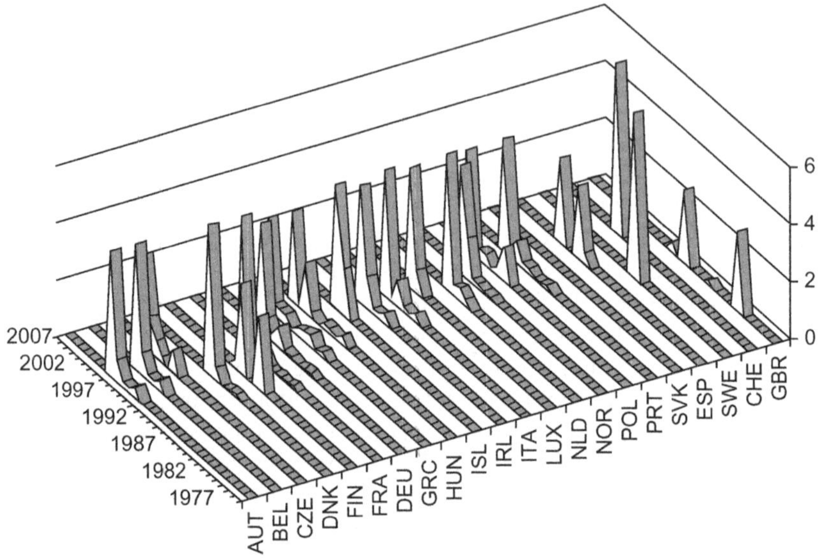

Fig. 1.1 Liberalisation waves in EU telecom markets, 1975–2007 (Nicita and Belloc 2016)

This process can be seen in Fig. 1.1 above showing the timing and intensity of telecommunications liberalisation across EU countries.[23] Due to European Commission directives and consequent transpositions into national legislation, at the end of the '90s, all European countries experienced a wave of progressive reductions in legal market entry impediments for alternative operators, hence, creating legal preconditions for an actual competition in the market.

As shown, the UK liberalisation was not the result of the European policy overhaul, as it was put in place at the beginning of the '80s with the UK Telecommunications Acts.[24] In this respect, liberalisation in the UK

[23]The liberalisation index is calculated as 1-year variations of a liberalisation index elaborated from OECD data on product market regulation. Specifically, it is obtained by subtracting the OECD's indicator of entry barriers to the telecommunications' market from its maximum value (the liberalisation index thus ranges from 0, minimum, to 6, maximum liberalisation)

[24]The 1981 UK Telecommunications Act liberalised terminals and value-added services, while the 1984 TA privatised the incumbent operator and introduced a duopoly by allowing entry of an alternative operator.

was rather a source of both political inspiration and a technical regulatory benchmark for EU institutions and other Member States.[25]

1.3 Substantial Liberalisation as Competitive Market Outcome

Liberalisation in the telecom market was introduced by EU policymakers in order to boost competition in downstream markets to the benefits of final consumers. However, the mere removal of exclusive rights was a necessary but not a sufficient condition to ensure an effective entry of new competitors, generating a full competition process.[26] The gap between a *formal liberalisation*—as a legislative outcome pursued through the removal of legal barriers—and a *substantial liberalisation*—as a market outcome comprised of competitive dynamics—emerges clearly in Fig. 1.2

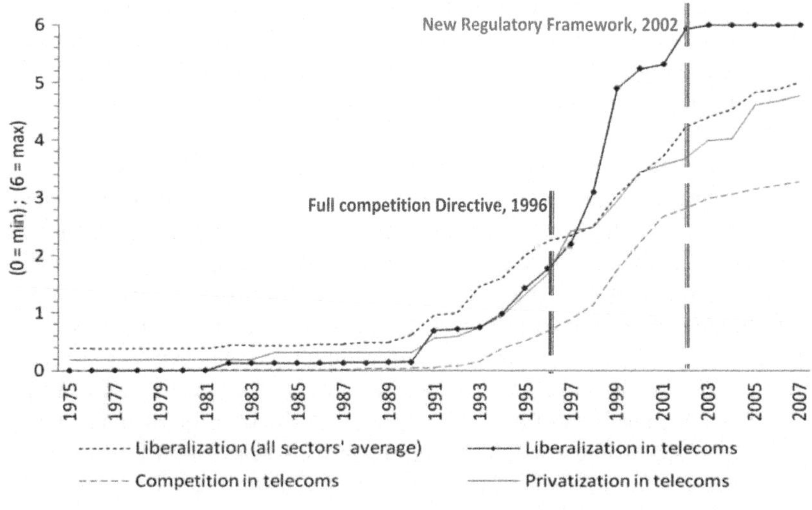

Fig. 1.2 The evolution of market-oriented policies, 1975–2007 (Nicita and Belloc 2016)

[25]The UK always has always been a laboratory for developing and defining new regulatory tools and methods and much of the EU legislation has been inspired to the UK approach. On the other side, often, EU law and case law enriched the UK regulatory and legal system with cross-sectoral consistency and harmonisation.

[26]Cave (2002).

below. The graph describes the intensity of liberalisation, privatisation and competition in the EU telecom markets as well as the average intensity of liberalisation in the other main EU network industries (electricity, passenger air transport, post, gas and railways).[27] The figure shows how the legal reforms abolishing entry barriers have been much stronger in the telecommunications sector than in the other network industries, especially after the full competition directive in 1996. However, the figure also reveals that in the same period the level of market competition, measured by the new entrants' shares, has tended to grow slower than the formal liberalisation process. After the adoption and national transpositions of the new regulatory framework in 2002, the gap between liberalisation and competition in telecoms has been slowly yet constantly reducing - also depending on the fact that formal liberalisation in 2002 has reached maximum intensity.

Indeed, introducing effective and sustainable competitive dynamics into a monopolised network industry is a much more complex and less immediate result. In particular, "successful liberalization is seldom as simple as removing all legal restrictions on entry into the regulated industry. Entrants face myriad economic barriers to entry, even when legal barriers are removed."[28]

Over the decades, telecom incumbent operators have developed ubiquitous physical networks. It was then extremely hard for any other operator to economically duplicate those networks that implies *high fixed costs.* Moreover, most part of fixed costs are *sunk costs*, that is those costs that cannot be recovered, creating an asymmetry of incentives between companies already operating in the market and others outside it: respectively a disincentive to exit and a barrier to entry. Finally, as we have seen, telecom markets are strongly impacted by *network externalities*, which may in turn generate another form of barrier to entry, due the high *switching costs* that incumbent's customers would sustain to move towards new entrants.

[27] Competition intensity is measured as a weighted average of the market share of new entrants in the trunk telephony market, in the international telephony market, and in the mobile market, and it is obtained by subtracting the OECD's indicator of incumbent's share from its maximum value (the competition index we use thus ranges from 0—minimum entrants' share— to 6—maximum entrants' share).

[28] Armstrong and Sappington (2006).

[29] Majone (1997).

The presence of a pervasive physical network, extensive network effects as well as economies of scale and scope, described in Sect. 1.1, all translates from an economic perspective into significant economic entry barriers and economic impediments for entrants to compete in the market effectively and sustainably. Therefore, even in a liberalised market, vertically integrated (state-owned) incumbents maintained a paramount competitive advantage. Consequently, an economic regulation is needed to address that competitive advantages, by allowing new operators to access, at competitive conditions, the incumbent's infrastructure, which is considered a competitive essential upstream input.

The role for the state within telecom markets governance evolved consistently with economic theory: no longer (only) related to a state direct intervention in the market but an indirect one, as an economic regulator. [29] Economic theory prescribes an intervention for public authorities aimed to identify and regulate *market failures*, namely, when market dynamics would naturally lead to sub-optimal resource allocation.[30] "The single most widely accepted rule for the governance of the regulated industries is regulate them in such a way as to produce the same results as would be produced by effective competition, if it were feasible". [31] Where entrenched market powers and strong network externalities are involved, public rules should attempt to re-establish a *workable competition* by counterbalancing the incumbent's dominance and actively preventing its potential abuses. When possible, a competitive result is better achieved indirectly, not by setting price and quantities at retail downstream level, but promoting the competitive process and focusing on upstream bottlenecks and dominance and then leave free (unregulated) competition in downstream markets. This policy strategy is called *pro-competitive regulation*.

Policy practice is to delegate this fundamental role to public independent authorities, as pro-competitive regulation should be enacted and enforced with a neutral approach. This is fundamental to ensure a fair and

[30] According to welfare economics standard results, a general economic equilibrium in a competitive market—where consumers and businesses maximise their utility functions—produces an allocation of resources for which it is impossible to increase the welfare of one individual without diminishing the welfare of another individual (Pareto efficiency—Allocative efficiency). Market failures are those situations whereby this welfare economics result does not hold, i.e., situations of market power, externalities, public good, asymmetric information and market incompleteness.

[31] Kahn (1970).

level playing field for all market players, especially in the case of liberalisation without privatisation, where vertically integrated state-owned firms face downstream competition (Sect. 1.5). This may sound trivial if we do not consider that most regulatory functions within monopolistic markets were carried out by the dominant firm itself. The OECD has issued increasingly clear guidelines regarding the regulators' independence in order to avoid any "undue influence".[32] In fact, formal liberalisation and pro-competitive independent regulation are in a complementarity relationship, as the pro-competitive impact of liberalisation is significantly higher when overseen by an ***independent regulatory authority***.[33] Accordingly, all sector-specific pieces of EU legislation for network industries[34] (including the telecom directives) have imposed on Member States strict requirements to ensure regulatory independence, not only from market players, but also from national governments and any other public body.

Therefore, as detailed in Chap. 2, the European institutions have complemented the liberalisation directives for the EU telecom markets with a harmonised regulatory framework hinging on pro-competitive independent regulation. The EU regulatory framework was established under the 2002 and 2009 directives, the so-called New Regulatory Framework (NRF),[35] recently updated and systematised by the 2018 EU Electronic Communications Code (EECC).[36] The NRF further enhanced the formal liberalisation by introducing a general authorisation regime instead of individual permissions to operate into the market. Moreover, a fundamental issue addressed by the new regime has been the maintenance of the universality of the service also in a liberalised and competitive context, by introducing ***universal service obligations*** (USO). Ideally, when competitive entry occurs, competition push companies to provide service over the whole market, at good quality and fair prices. However, in a liberalised

[32] OECD (2016); OECD (2017b).

[33] Belloc et al. (2013).

[34] The main acts regulating utilities and network industries all have provisions related to regulator independence, i.e., Telecommunications (Directive 2018/1972 establishing the European Electronic Communications Code), Gas (Directive 2009/73/EC concerning common rules for the internal market in natural gas), Electricity (Directive 2009/72/EC) and Railways (Directive 2012/34/EU establishing a single European area).

[35] Directive 2002/21/EC on a common regulatory framework for electronic communications networks and services (Framework Directive—FD).

[36] Directive (EU) 2018/1972 establishing the European Electronic Communications Code (Recast).

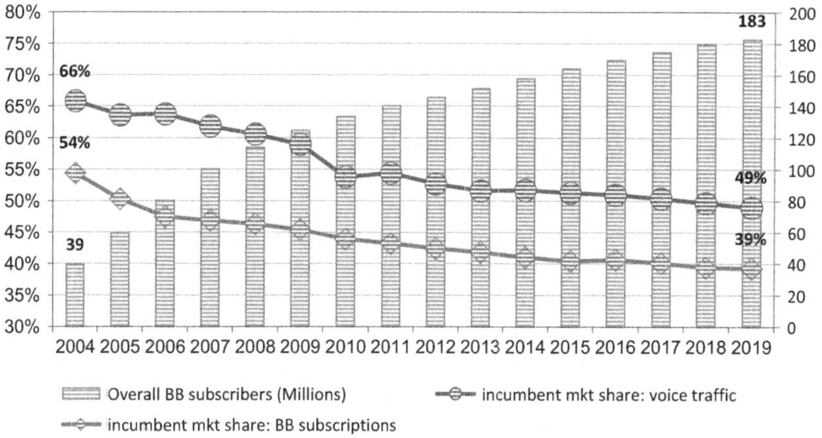

Fig. 1.3 Incumbents average market share and overall broadband subscriptions in EU

context a "safety net" is necessary to ensure that a set of basic services is available to all end-users at an affordable price, even in those areas that are not commercially profitable and a liberalised market would not serve. The lack of those services would otherwise prevent citizens from full social and economic participation in society.[37]

As mentioned, the gap between competition and liberalisation intensity decreased more than proportionally after the adoption of the new regulatory regime (Fig. 1.2). Indeed, one simple way to observe and quantify the impact of liberalisation and pro-competitive regulation is by looking at the trend of incumbent market share and, hence, to what extent competitors have been able to enter the market and sustainably compete. Looking at the average incumbent market shares in voice fixed telephony (the traditional legacy service for the sector) and broadband subscription (the current core service), a clear declining trend can be observed (Fig. 1.3).[38] Incumbent market shares for fixed voice traffic volume dropped from 65% in 2004 to about 49% in 2019, as did the market shares in broadband (BB) subscription, which decreased by 15 percentage points in 15 years, getting to 39% in 2019. These trends clearly signal the effective impact of liberalisation and pro-competitive regulation in the EU.

[37] Recital 212 EECC.

[38] Data for Figs. 1.3, 1.4, 1.5 and 1.6 comes from European Commission (2019a, 2019b).

The constant rise in demand, especially for innovative products and services, such as EU BB subscriptions (from 39 to 189 million in 15 years), is another important by-product of market liberalisation and competition enhancement. It reveals how much liberalisation and pro-competitive access regulation have spurred technological innovation across the market and oriented productions consistently with consumers' preferences and needs, sustaining their willingness to pay.

A similar trend can be observed in mobile telephony, although mobile markets are genetically different from those of fixed telecom as their development has taken place more recently without decades of legal monopolisation. However, mobile markets are subject to technical and economic constraints arising from the scarcity of radio-spectrum frequency resources, which has pushed policymakers to establish a licensing regime, consequently resulting in a tight oligopoly market structure. Due to a more balanced initial market structure and in consideration of the continuous vigorous expansion of demand in terms of subscribers (from 387 million in 2004 to 701 million in 2017), market dynamics in the mobile industry have been more fluid, as well as more heterogeneous. EU national leading mobile network operators (MNOs) market share slightly declined from 40% in 2004 to 35% in 2017, although it was not as constant as in the fixed voice and BB markets (Fig. 1.4).

In facts, the declining trends of EU leading mobile network operators have been reverted in the last few years, also due to a generalised EU-wide (and worldwide) wave of consolidation in the mobile market, particularly through mergers and acquisitions.[39] On the other side, the resulting tight oligopoly structure has been softened by the entrance of mobile virtual network operators (MVNOs), which are competing players accessing and

[39] Europe has experienced a wave of mergers in the mobile telecom markets leading to a consolidation of the tight oligopoly market structure. The EU Commission has recently allowed mergers restricting from 4 to 3 the number of MNOs in the Netherlands, Austria, Ireland, and Germany, but rejecting similar mergers in Denmark in 2015, in the UK in 2016, and, in Italy, another proposed merger has been approved subject to divestiture remedy and the consequent entrance of another operator. Earlier decisions had dealt with, and approved, 5-to-4 mergers in Austria, the Netherlands, and the UK. See Genakos et al. (2018).

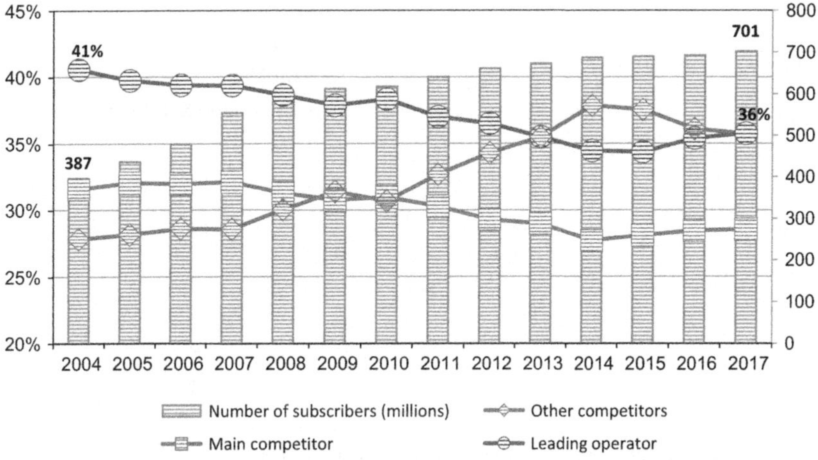

Fig. 1.4 MNOs average market share (subscribers) and overall subscribers in EU

using MNO's networks.[40] Overall, MNVO average market share in terms of subscriptions in EU amounted at around 10% in 2019.

In addition, it is meaningful to look at the downward trend of prices, which are usually inversely related to competition intensity in different market segments, and sustain the increase in demand in competitive markets. The comprehensive Eurostat index for "telephone and telefax services" has fallen from 12 percentage points between 2006 and 2015.[41] Moreover, Fig. 1.5 also describes the much sharper decreasing evolution of prices for specific services, subject to increasing competitive pressure: (a) fixed broadband services (8 < Mbps < 12) and voice, and (b) ARPU for mobile services.

[40] MVNOs have own customer management activities and may issue their own SIM cards and have some mobile network switching structures, but, in any event, no transmission structures and spectrum frequency resources assigned. Differently from fixed telephony, due to the competition based on different networks—although limited to a few operators—in most cases MNOs are not ex-ante obliged to give access to third parties (need to show joint dominance and more than "normal" prices and profits—only Spain in 2006), so access takes place on the basis of commercial negotiations and agreements. In some circumstances, MVNO entry was a consequence of antitrust commitments, as for the 2007 Italian case A357 Tele2/Tim-Vodafone-Wind.

[41] Eurostat database: https://ec.europa.eu/eurostat/web/hicp/data/database

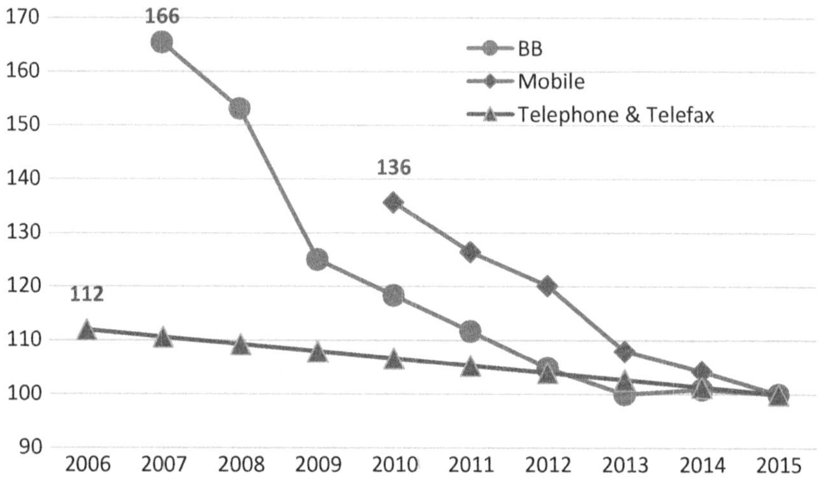

Fig. 1.5 Price trends in telecom markets (base 100 in 2015)

The impact on demand and prices, as well as on innovation, can be considered indicators of the impact of liberalisation and pro-competitive regulation on social welfare. Moreover, price and demand trends of innovative products, as BB and ultra BB, signal how much liberalisation and pro-competitive access regulation have influenced technological innovation of the companies operating at the different stages of the market value chain.

1.4 Regulatory Failures and Other "Policy Variables"

The development of a competitive market structure depends on formal liberalisation and pro-competitive regulation, as well as on how single regulators enforce the existing regulatory framework and consequently how often *regulatory failures* may occur.

Regulatory failure refers to a situation where regulation intervenes either unnecessarily or to address a market failure but fails and ultimately produces additional inefficiencies in market functioning.[42] As mentioned,

[42] Buchanan (1988).

the desirability of public intervention for an economic regulation (thus not a social regulation aimed at achieving socially equitable redistributive outcomes) is linked to the resolution of ***market failures***. Only in these contexts economic regulation benefits can outweigh its costs.

Moreover, the authoritative determination of behavioural rules, based on a centralisation of economic decisions, is inevitably subject to the problem of finding the information necessary for the definition of socially optimal private behaviours. In a physiological competitive market process, this information, private and dispersed among market players, is made publicly available through the price, while the extraction of information carried out by public authorities has a cost and tends to be imperfect. Therefore, information available to regulators often happens to be insufficient, causing strong informational asymmetries vis-à-vis regulated companies (which may further have strategic incentives to conceal or mis-report information). For all these reasons, regulatory intervention must remedy market failures preserving as far as possible the decentralisation of economic decisions, complying with the ***proportionality principle***, that is by effectively addressing the problem at stake while imposing the smallest restriction possible on market players' freedom.

Furthermore, public intervention always runs the risk of diverting from public interest, to pursue private interests, such as those of the recipients of the intervention, the so-called ***regulatory capture***, or that of the public apparatus itself.[43] In addition, according to "what level of public institutions the public interest is defined at", i.e., at local vs national vs EU level, the pursuit of specific local or national interests, opposing to others, may negatively affect the overall EU social welfare and consolidation of a EU-wide internal market. Consequently, as specified in Chap. 2, a consistent and harmonised enforcement of the EU framework at national level, implemented via coordination procedures involving the EU commission and all national regulatory authorities, is also aimed at minimising regulatory failures and institutional externalities.

Nevertheless, despite the harmonised pro-competitive regulatory framework, the declining trend of incumbent market shares in the liberalised fixed telecom markets varies considerably from country to country in Europe. In 2017, incumbent in Ireland held 32% market share in fixed

[43] In this sense, the descriptive theory of private interest, as opposed to the prescriptive theory of public interest. See, Stigler (1971) as for regulatory capture; and Niskanen (1968) as for incentives of bureaucracy.

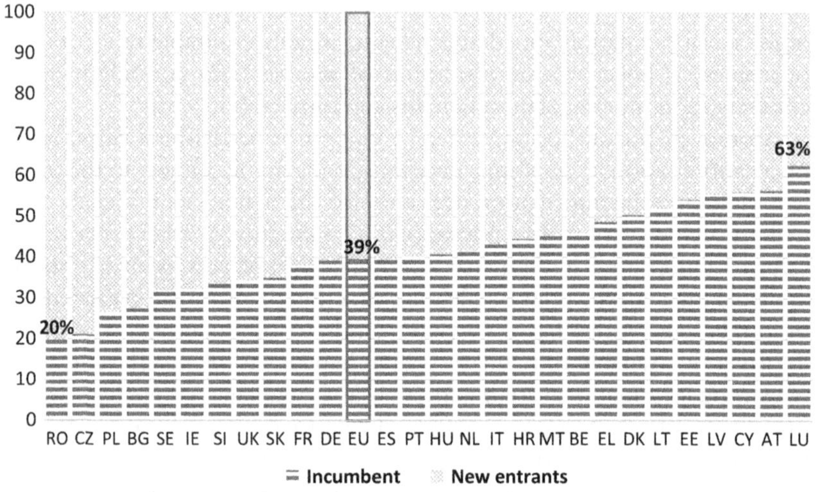

Fig. 1.6 Incumbent market shares in fixed Broadband, per MS in 2019

telephony versus 90% in Lithuania. As for broadband services, in 2019 Romanian incumbent held 20% market share, whereas 63% of incumbent market share in Luxemburg (Fig. 1.6).

These divergent outcomes can of course be explained by different national market characteristics and economic strategies of private companies. However, other crucial factors related to public intervention - *policy variables* - must be considered as, admittedly, they can contribute to explaining the existing market shares differences, even under a homogeneous regulatory framework and a consolidating EU internal market.

A first general consideration concerns the relationship between economic regulation and competition law enforcement, which have been traditionally considered as two distinct policy tools aimed to pursue public interest, i.e., competitive market dynamics and maximisation of social welfare. Based on this, ineffective, incomplete, or unbalanced regulatory measures have often requested regulation to be supplemented by competition law enforcement, resulting in an interplay of pro-competitive public policies. How the relationship between these policies is designed and implemented (as complements as in the EU, or as substitutes as in the US) may have an important impact on the evolution of the competitive dynamics and outcome in the market (Chap. 3). Although the interplay of competition law enforcement and economic regulation is harmonised at EU level, the interaction and coordination between national regulatory authorities, competition authorities and national courts are very much based and steered by national legislation, case-law and administrative practice. This aspect represents an additional policy variable which can profoundly affect the pro-competitive market outcome in a liberalised and regulated sector.

Another crucial variable within market liberalisation are demand-side obstacles to competition, which have been neglected in the initial design of liberalisation policies, focusing solely on the supply side impediments. As described in Chap. 4, demand-side policies revealed to be crucial for complementing pro-competitive regulation (and competition law enforcement) in order to obtain a substantial liberalisation outcome. Indeed, policy actions aimed at empowering consumers, as demand-side market players, are key to the efficient market functioning, and not only to protection of consumers. Here, reducing switching costs faced by consumers and preventing lock-in phenomena should be regarded as pro-competitive regulatory measures. In the same vein, regulatory actions focused on fostering consumer market trust and service transparency are of the utmost importance from an economic perspective.

The European approach has always placed a great emphasis and attention on consumer protection issues.[44] Nevertheless, consumer empowerment policies may not be completely effective if actual consumer thinking, decision-making and behaviour—such as consumer inertia and other *cognitive biases* [45]—are not duly targeted by regulation. Switching rates are still quite low, especially for fixed telephone services because regulatory measures designed around a behavioural approach have still not been extensively implemented within the EU legislation.[46] In actual fact, behavioural aspects are mainly left to national legislation and regulator initiatives. Consequently, the different national regulatory cultures and competences may lead to divergent approaches to demand-side remedies

[44] See Weatherill (2013); Micklitz et al. (2009).

[45] Deviating from the classical assumption of consumers as fully rational agents, which has been contested since the fifties. See Simon (1955).

[46] Jolls (2017); Zamir and Teichman (2017).

across the Internal Market, additionally explaining the heterogeneity of regulatory outcomes across EU national markets.

Furthermore, a complex aspect of liberalisation in a technological and investment-intensive market, such as the telecom industry, concerns the relationship between the goal of nurturing competition and the need to ensure incentives for innovation and investments. The policy debate on this topic has been rather intense. Within the main current economic narrative, the rivalry between competing companies displays a downward pressure on prices and an upward influence on quality, and also—in its dynamic aspect, i.e., *dynamic competition*—it is a driver for innovation and investment in new technology and services. On the other hand, it is argued that (excessive) market competition is detrimental to innovation and investments, which would be better pursued in a monopolistic or oligopolistic context. Both views have merit and rest on economic theory pillars (as in Schumpeter and Arrow theories[47]). Nevertheless, this theoretical dichotomy outlines the existing difficult trade-off between competition enhancement and innovation/investment promotion, which regulators have been called upon to disentangle.

Competition principle was at the basis of the liberalisation process and was defined as one of the main objectives of the 2002 EU telecom directives. Promotion of investments was considered as a tool to promote a long-term sustainable and durable competition. Consistently, this potential trade-off has been framed as an interplay between two possible types of competition. On the one hand, a competition based on downstream services, where new entrants completely rely on incumbent infrastructure and, on the other, a more challenging (both for companies and regulators) market context, where competitors develop their own infrastructure by creating a more intense and sustainable competition between alternative vertically integrated operators. The approach adopted, known as *ladder of investments*, attempts to strike a balance in the long run regarding these different goals. (Chap. 5)

Under this balanced approach, technological innovation and innovative investments are sustained and pursued by both competitive dynamics and incentive-based regulation.[48] Nevertheless, very advanced technological

[47] Arrow (1962); Schumpeter (1942).

[48] For incentive-based regulation see Sect. 5.3.

innovations have the potential to disrupt existing market strategies and dynamics, thus resulting in a material impact on market outcome and liberalisation effectiveness.[49] In this light, recent technological innovation in the telecom sector (such as the deployment of fibre networks and mobile broadband services and networks evolution) have deeply impacted the telecom markets. Therefore, public bodies, both policymakers and regulators, have had to reconsider and recast their policies and strategies.

Fibre networks are a material improvement for broadband connectivity and the sustainable development of innovative services. Accordingly, the development of fibre networks (or, more generally, *Very High Capacity Networks*, as defined by the EECC) has become one of the main objectives of the telecom regulatory framework. Against this background, a major disruption arises as a challenging problem: the capital-intensive investments necessary to deploy fibre networks, according to the most recent European and national policy targets,[50] risk being unachievable, if a new regulatory balance between competition and innovative investments is not found. However, this in turn may imply an undesirable (and unintended) softening of the push that competition exerts toward dynamic efficiency.

So far, responses have emerged both from the market and from policymakers with new business models (e.g. wholesale-only operators) and market strategy (i.e., cooperative investments and network sharing) being developed by the telecom companies to navigate the new market conditions. Moreover, the regulatory framework has required regulators to weigh their pro-competitive measures (i.e., access regulation) so as not to inhibit investments and VHC network deployment. Finally, in most EU countries, public direct interventions aimed at financing the most innovative investments have been established and need now to be reconciled with existing state aid rules and competitive dynamics. (Chap. 6)

Additional elements of innovation in the telecom industry are related to the evolution of mobile communication and, namely, mobile broadband. Over the last five years, there has been a fast expansion of internet

[49] See OECD (2017a).

[50] Digital Agenda for Europe (defining targets for 2020) and Gigabit Society (defining targets 2025).

services consumed in mobility by people. On the other hand, extensive Machine-to-Machine (M2M) communications, also in the context of the so-called Internet of Things (IoT), have begun to make up an important share of all mobile communications. 5G, the latest mobile communications technology standard to enter the market, represents a new way of providing mobile ultra-high capacity and very low latency connections between people and things, supporting a wide range of innovative applications and services (e.g. assisted driving, home and industry automation, eHealth, energy management, etc.). These phenomena are heavily contributing to the widespread use of mobile data traffic and are bringing about a paradigm shift in the existing economic patterns and rationales of mobile markets. All this has raised the need for an update of some regulatory measures, and above all have called for a deeper rethinking about spectrum policy strategy, and for the adoption of a more systemic approach. (Chap. 7).

Finally, a major innovation within the telecom industry, and the most profound, concerns the digital transformation of the sector and of the entire economy. Digitalisation impacts all sectors horizontally, nevertheless, telecom as well as media markets are at the forefront. Firstly, digitalisation implied a process of *multimedia convergence* allowing for a range of different digital content and services to be transmitted on the same digital network. Secondly, the *decoupling* between networks and services has entailed a modularity of the digital ecosystem which supported the emergence of global on-line platforms, offering content and services *over-the-top* (OTT) of the traditional telecom value chain. These actors have dramatically influenced market dynamics in telecom and media markets, drawing a great deal of attention from the public, policymakers and stakeholders. Crucial topics concerning the relationship between OTTs and traditional players, such as a regulatory playing field and *network neutrality*, have been at the core of the policy debate for the last years, and will, likewise, also animate it over the next few years (Chap. 8). In a sense, global platforms are not a policy variable, but a constant, as all world countries are impacted. Hence, it is still influencing the liberalisation outcome, not increasing differentiation but, on the contrary, pushing toward the development of consistent and increasingly harmonised digital and electronic communications policies.

Box 1.1 The EU Digital Strategies

Digital transformation and platform economy originated horizontal multi-faceted policy issues with a global dimension. The development of differentiated and fragmented national digital public policies is a non-trivial risk. Consistently, the digital transformation has been approached by the European Commission launching periodic Digital Strategies based on the internal market objective and the subsidiarity principle.

The Digital Agenda for Europe (DAE) [51] adopted in 2010 was one of the 7 flagship initiatives of the Commission's Europe 2020 Strategy for smart, sustainable and inclusive growth. This policy programme has attempted to coordinate at EU level the public policies to ensure access to online activities for individuals and businesses under conditions of fair competition, of a high level of consumer and personal data protection, irrespective of their nationality or place of residence.

The Digital Single Market Strategy (DSM)[52] lunched in 2015 is made up of a few specific initiatives, aiming to tackle the major obstacles to the development of EU cross-border e-commerce. The three pillars of the EU DSM strategy are:

(a) Providing a better access for consumers and businesses to online goods and services across Europe, e.g., by facilitating cross-border e-commerce, by limiting unjustified geo-blocking practices, by modernising the EU copyright framework (cross-border content access), and by protecting online consumer rights.
(b) Creating the right conditions and a level playing field for digital networks and services to flourish by: ensuring generalised connectivity, investment incentives for very high-capacity networks and accelerating 5G wireless roll-out; reviewing the AVMS directive and adapting the existing rules to new models for content distribution; reinforcing trust and security in

(*continued*)

[51] European Commission (2010). This was one of the 7 flagship initiatives of the Commission's Europe 2020 Strategy for smart, sustainable, and inclusive growth.

[52] European Commission (2015).

Box 1.1 (continued)

digital services (cybersecurity) and in the handling of personal data (digital privacy); defining a transparent, open and non-discriminatory regulatory context for online platforms, including the removal of illegal online contents.

(c) Maximising the growth potential of the European Digital Economy by: addressing the barriers in the free flow of non-personal data in order to boost the data economy; focusing on standards and interoperability; supporting an inclusive digital society through the development ICT professionalism and public e-government plans

In April 2020, the Commission lunched the strategy "Shaping Europe's digital future,"[53] where are laid out the main pillars of next five-year policy strategy:

(a) Promotion of new adequate investments at EU, national and regional levels in VHCN, 5G networks, and "deep tech" to develop and use digital solutions at scale and to strive for interoperability in key digital infrastructures
(b) Legislative framework for data governance, facilitating forms of data control and data sharing within the Internal Market.
(c) New Consumer Agenda aimed to empower consumer to make informed choices and play an active role in the digital transformation.
(d) Focus on platform economy by (d_i) ensuring that global on-line platforms market power is not harmful for fairness and openness of European markets, and (d_{ii}) addressing with strengthened and modernised ex-ante rules (tax, consumer protection, etc.) all other systemic problems that may arise for digital services across the EU.

[53] European Commission (2020).

1.5 Liberalisation and Privatisation: Differences and Interdependences

Liberalisation and its outcomes may also be influenced by privatisation policy, which might be an additional element in the design of market-oriented policy. Privatisation concerns the shift from public to private ownership or control of the (former) monopolist. Indeed, an important debate has revolved around the interdependence of liberalisation and privatisation policies, in particular with reference to their sequencing and intensity,[54] and whether privatisation of the incumbent firms could have an additional pro-competitive impact onto the liberalisation process.

Considering the different possible combinations of liberalisation and privatisation, it is useful to present a (static) taxonomy of market-oriented policy outcomes (Table 1.1).

The diagram above can be read by starting from a situation of a state-owned monopoly (quadrant I). Liberalisation, as previously described, refers to the reduction of legal market entry barriers, allowing new private companies to compete downstream with the state-owned (vertically integrated) firm (quadrant III). Privatisation refers instead to the elimination

Table 1.1 A taxonomy of market-oriented policies

Privatisation \ Liberalisation	NO	YES
NO	(I) Monopolistic State-owned firm	(III) (vertically integrated) State-owned firm and downstream competition
YES	(II) Monopolistic Private firm	(IV) (vertically integrated) State-owned firm and downstream competition

[54] Nicita and Belloc (2016).

or reduction of public institution ownership and/or control of the company, resulting in a market structure with a private monopolistic firm (quadrant II). Of course, both market-oriented policies could be jointly adopted, in this case, the outcome will be a (vertically integrated) private firm facing new entrant downstream competition (quadrant IV).

According to the mainstream economic narrative, "there is one overriding advantage in private provision: the private sector is usually more efficient. Because private operators focus on profits, they have strong incentives to reduce their costs. This is true even for private monopolies as compared to public ones. (But of course, consumers do not benefit from these efficiencies unless the lower costs are somehow passed on to them.) Private providers also have incentives to innovate, which benefit consumers."[55] Moreover, "the government's objectives are multiple and fuzzy and can change over time; this exacerbates the problem of managerial control in public enterprises. [...] A government must prevent monopoly pricing, control quality, reduce negative externalities, encourage sectoral policies and national independence, and concentrate on investment and employment in recessions. Thus, government ownership of firms is problematic when defining the goals of the firms. [...] the problem with many government objectives is that, unlike profit maximization, they are hard to contract upon."[56]

In other words, public-owned enterprises are considered inefficient, primarily because their management (a) usually do not have clear incentives to improve productive efficiency as they have no pressure from shareholders, and (b) may pursue some political objectives (imposed by politicians) negatively impacting the company productive efficiency, and resulting, for example, in excessive or unbalanced employment, or inefficient acquisitions of failing companies.[57]

However, there is no general consensus on this approach. For example, other economists argue that (large) private firms could be subject to the

[55] OECD (2014).

[56] Roland (2008).

[57] Boycko et al. (1996) A Theory of Privatisation. The Economic Journal, 106(435), 309-319; Armstrong and Sappington (2006). Capobianco and Christiansen (2011) Competitive Neutrality and State-Owned Enterprises: Challenges and Policy Options", OECD Corporate Governance Working Papers, No. 1, OECD Publishing; Megginson and Netter (2001) From State to Market: A Survey of Empirical Studies on Privatization. Journal of Economic Literature, 39 (2):321-389

same corporate governance problems affecting public management (i.e., principal-agent relationship and opportunism, asymmetric information, and conflicts of interest). Moreover, privatisation without liberalisation (quadrant II) may generate a much greater risk of anti-competitive behaviours as a private monopolistic firm, pursuing profit maximisation, "is better at exploiting monopoly power than the government."[58] According to some, the real comparative advantage of privatisation over public ownership and management is to be found in the role that property rights play in increasing incentives to invest when contracts are incomplete.[59] Under incomplete contracting, ex-post decisions in unforeseeable states of the world are taken by the agent who has the residual rights to control proprietary assets. Thus, it is in the interest of the owner to make (specific) investments that maximise the value of the assets.[60]

On the opposite side of the taxonomy, in a scenario of liberalisation without privatisation (quadrant III), a state-owned firm competes in the downstream market against new entrants. Downstream competition would provide incentives to set competitive downstream prices (on top of pro-competitively regulated upstream prices), and the monopoly deadweight loss would be reduced. Moreover, incentive-based regulation would push the incumbent to produce the upstream input in an efficient way. In the absence of effective regulation by an independent authority, however, a state-owned incumbent may still have an unfair competitive advantage (greater than a dominant private company) due to indirect support that government and other public institutions could provide. According to the European Commission, "the potential problems associated with public ownership include not only control issues but, potentially, also preferential access to capital and distortions of competition both for and in the market, which can be exacerbated by regulatory capture."[61] Whereas, in a liberalised market subject to independent regulation, most of the competitive problems are not related to the public or private ownership of the former monopolist.

European institutions, as well as independent regulatory and competition authorities, have constantly kept under surveillance the relationship between national Member States and public-owned enterprises in order to

[58] Stiglitz (2008).
[59] Laffont (2000).
[60] Nicita and Pagano (2016).
[61] European Commission (2013).

maintain a *level playing field* and fair competition in liberalised markets. The enforcement of the state aid rules coupled with the independence of national regulatory authorities from national governments have traditionally been the main tools to reach this goal.[62] Nonetheless, it is important to stress that European Union institutions have never encouraged or mandated the privatisation of state-owned enterprises competing in liberalised markets. Such an agnostic approach by the European Union, guaranteed by art. 345 of TFEU, is due to the facts that (a) there is no clear consensus in economic theory that privatisations have a positive impact on social welfare in all market and social contexts and (b) ownership and property rights have always been considered by EU institutions a very sensible policy domain to be dealt with at national level.

Nevertheless, in order to provide an additional market-oriented direction and to improve public finance, some forms of privatisation have been put in place to a different extent in different European countries. Some telecom incumbents were completely privatised (such as in the UK, Denmark, Italy, Spain, the Netherlands, Portugal), whereas, in quite a few countries, public shareholders are still active, sometimes, holding control (such as in Austria, Belgium, Germany, Greece, France, Romania, Lithuania, Poland, Sweden and Slovenia).[63]

Some scholars have argued that "competition is difficult to achieve within the public sector, so that there is a natural complementarity between liberalization and privatization".[64] When both liberalisation and privatisation policies are implemented, a former vertically integrated public monopoly gives way to a competitive downstream market, where private firms compete by enjoying access to the infrastructure on an equal basis (quadrant IV). In this model, the principal-agent problems associated with both public and private ownership and the monopoly deadweight loss are curbed by market forces. Hostile takeovers can further align managers' incentives to run the firm under profitability criteria. Thus, "introducing competition into previously monopolised and regulated network

[62] Here, the EU prescription of independent regulation makes it possible to efficiently separate liberalisation and privatisation policies.

[63] As of 2016: Austria (28.4%), Belgium (55%), Cyprus (100%), Germany (32%), Greece (10%), France (26%), Romania (45% plus 100% of another operator), Lithuania (100% of an alternative operator), Poland (68% of 2 alternative operators), Sweden (37%) and Slovenia (62.5%)

[64] D. Newbery (1997).

utilities is the key to achieving the full benefits of privatisation. Privatisation is necessary but not sufficient."[65]

However, the OECD also underlines that privatisation is not always necessary nor desirable. In specific situations, private provision may have a negative impact on social welfare, as "there may be situations in which a private provider can reduce costs by reducing quality, which harms consumers, and, importantly, government oversight cannot remedy the situation. It may not be possible for the government to observe and measure reductions in quality. In this situation, if competition is not feasible or is ineffective, if the opportunity for innovation is limited, and if gaining a reputation as an efficient service provider is unimportant in the sector, continued public provision could be preferable."[66] Admittedly, when it comes to chronic lack of competition, systemic positive social externalities and deep asymmetric information, private ownership may result in increasing market failures.

For these reasons, and because of the financial crisis impact, in some national contexts where privatisation has taken place, there has been a growing political debate about new public direct interventions in the market for services of general economic interest. Those advocating for public control of telecom companies stress two arguments. First, only the public sector would have the financial resources to build new networks with a comprehensive, if not complete, coverage of the territory and population, and second, the (national) security of networks has become so important for society that it should be managed by the state.

This policy debate on re-nationalisation is often accompanied by considerations regarding the re-monopolisation of the upstream market, which would lead to a substantive counter-liberalisation wave. As in a circular policy cycle, a new monopolisation, limited to the upstream wholesale market segments (see Fig. 2.1), would strive to avoid inefficient duplication of extremely costly investments in new networks (an update of natural monopoly argument) and avoid coordination costs and strategic behaviour which would be otherwise necessary to grant the security of a number of different networks (a national interest rationale). EU Member States have strong legal constraints regarding re-monopolisation. In actual

[65] D. Newbery (1997).
[66] OECD (2014).

fact, the tools and strategies considered do not entail new special and exclusive rights but, rather, the "facilitation" of market concentration through mergers and acquisitions, sometimes linked to a vertical (voluntary) separation of the incumbent as a remedy to the significant reduction in competition (see Sect. 2.3 and Sect. 6.2).

This scenario is sometimes further motivated as a response to challenges arising from digital platforms and the current decline of telecom companies' profitability and competitivity. In the same vein, some scholars and experts argued that, after sector liberalisation, the European regulation negatively shaped the telecom industry structure by allowing too much entry with too many operators, active at national level, and strongly competing in a "price war", especially in the mobile sector, at the expenses of innovation and investments. Under this respect, consolidation through mergers has been often suggested as a successful way to restore a balanced equilibrium, as the one observed, for instance, in the United States.

Consistently, over recent years, there has been a wave of mergers and acquisitions, particularly in the mobile telecommunications. As a matter of fact, addressing fragmentation of telecom markets is one of the pillars of the DSM strategy, however this "has been interpreted differently by different stakeholders. For the Commission, fragmentation relates to fact that the availability, quality, and prices of services vary significantly across the continent, in telecoms markets defined by national borders. Mobile operators, instead, point to the fact that there are about 40 mobile network operators in the EU. Many operate in just one or two countries. By comparison, in the US there are four nationwide mobile operators (AT&T, Verizon, Sprint, and T-Mobile). While the Commission as the EU's competition authority might be lenient in cross-border merger cases, the mobile operators appear more interested in achieving within-country consolidation."[67]

On the other side, other scholars instead argued that in the US a decline in competition has been brought about in large part by the rise of very powerful technology companies through a light-handed antitrust enforcement and regulatory oversight.[68] In a similar vein, others have criticised the US regulatory and antitrust approach in the telecom industry, as it allowed few massive cable and telephone companies to dominate the US

[67] Cave et al. (2019).
[68] Philippon (2019).

telecommunications, increasing prices and also negatively affecting the level of consumer protection, which is much lower than in Europe. [69]

Indeed, there is an extremely complex evolutionary relationship between all "policy variables" composing the liberalisation outcomes, including policies about direct public control and consolidation. The evolution of the governance of telecom markets has been the result of such complexity and this is why only by fully considering and understanding all variables involved, it is possible to learn important lessons on the coevolution of markets, rules and public institutions and ultimately try to apply those lessons to govern current and forthcoming challenges.

References

Amato, G., & Laudati, L. (2001). *The Anticompetitive Impact of Regulation*. Northampton: Edward Elgar.

Armstrong, M., & Sappington, D. (2006). Regulation, Competition and Liberalisation. *Journal of Economic Literature, 32*(2), 353–380.

Arrow, K. (1962). Economic Welfare and the Allocation of Resources for Invention. In: The Rate and Direction of Economic Activities.

Baumol, W. (1977). On the Proper Cost Tests for Natural Monopoly in a Multiproduct Industry. *The American Economic Review, 67*(5), 809–822.

Belloc, F., Nicita, A., & Parcu, P. (2013). Liberalizing Telecommunications in Europe: Path-dependency and Institutional Complementarities. *Journal of European Public Policy, 20*(1), 132–154.

Boycko, M., Shleifer, A., & Vishny, R. (1996). A Theory of Privatisation. *The Economic Journal, 106*(435), 309–319.

Buchanan, J. (1988). Market Failure and Political Failure. *Cato Journal, 8*, 1.

Capobianco, A., & Christiansen, H. (2011). Competitive Neutrality and State-Owned Enterprises: Challenges and Policy Options", OECD Corporate Governance Working Papers, No. 1.

Cave, M. (2002). Barriers to Entry in European Telecommunications Market. In J. Jordana (Ed.), *Governing Telecommunications and the New Information Society in Europe* (pp. 47–65). Cheltenham: Edward Elgar.

Cave, M., Genakos, C., & Valletti, T. (2019). The European Framework for Regulating Telecommunications: A 25-year Appraisal. *Review of Industrial Organization, 55*, 47–62.

Crawford, S. (2013). *Captive Audience: The Telecom Industry and Monopoly in the New Guilded Age*. New Haven: Yale University Press.

[69] Crawford (2013).

Czernich, N., Falck, O., Kretschmer, T., & Woessmann, L. (2011). Broadband Infrastructure and Economic Growth. *The Economic Journal, 121*(552), 505–532.

European Commission. (1987). Green Paper on the Development of the Common Market for Telecommunications Services and Equipment—COM (87)290FINAL.

European Commission. (1999). Communication: Toward a New Framework for Electronic Communications Infrastructure and Associated Services—The 1999 review—COM (199) 539 final.

European Commission. (2010). Communication: A Digital Agenda for Europe—COM (2010)0245.

European Commission. (2013). Market Functioning in Network Industries: Electronic Communications, Energy and Transport, European Economy, Occasional Papers 129.

European Commission. (2015). Communication: A Digital Single Market Strategy for Europe—COM (2015) 92 final.

European Commission. (2019a). Digital Agenda Scoreboard.

European Commission. (2019b). Digital Economy and Society Index (DESI)—Connectivity Report.

European Commission. (2020). Communication: Shaping Europe's digital future—COM (2020) 67.

Genakos, C., Valletti, T., & Verboven, F. (2018). Evaluating Market Consolidation in Mobile Communications. *Economic Policy, 33*, 45–100.

Hellwig, M. (2009). Competition Policy and Sector-specific Regulation for Network Industries. In X. Vives (Ed.), *Competition Policy in the EU*. Oxford: Oxford University Press.

Jolls, C. (2017). Bounded Rationality, Behavioral Economics, and the Law. In F. Parisi (Ed.), *The Oxford Handbook of Law and Economics: Volume 1: Methodology and Concepts*. Oxford: Oxford University Press.

Kahn, A. (1970). *The Economics of Regulation: Principles and Institutions*. Cambridge: MIT press.

Laffont, J. J. (2000). *Incentives and Political Economy (Clarendon Lectures in Economics)*. Oxford: Oxford University Press.

Leibenstein, H. (1966). Allocative Efficiency vs 'X-Efficiency'. *American Economic Review, 56*, 392.

Majone, G. D. (1997). From the Positive to the Regulatory State: Causes and Consequences of Changes in the Mode of Governance. *Journal of Public Policy, 17*, 139–167.

Megginson, W., & Netter, J. (2001). From State to Market: A Survey of Empirical Studies on Privatization. *Journal of Economic Literature, 39*(2), 321–389.

Micklitz, H. W., Reich, N., & Rott, P. (2009). Understanding EU Consumer Law, Intersentia.

Napolitano, G. (2005). Towards a European Legal Order for Services of General Economic Interest. *European Public Law, 2005*, 566.

Newbery, D. (1997). Privatisation and Liberalisation of Network Utilities. *European Economic Review, 41*(3–5), 357–383.

Nicita, A., & Belloc, F. (2016). Liberalizations in Network Industries. Economics, Policy and Politics. Springer

Nicita, A., & Pagano, U. (2016). Finance-technology Complementarities: An Organisational Equilibrium Approach. *Structural Change and Economic Dynamics, 37*, 43–51.

Niskanen, W. A. (1968). The Peculiar Economics of Bureaucracy. *American Economic Review, 2*, 293.

OECD. (2000). Regulatory Reform in Network Industries: Past Experience and Current Issues. *OECD Econ Outlook, 67*, 151–171.

OECD. (2014). Economic Policy Reforms, Going for Growth.

OECD. (2017a). Key Points of the Hearing on Disruptive Innovation.

Philippon, T. (2019). *The Great Reversal: How America Gave Up on Free Markets.* Harvard University: Press.

Posner, R. (1975). The Social Costs of Monopoly and Regulation. *Journal of Political Economy, 83*, 807–827.

Roland, G. (2008). *Privatisation: Successes and Failures.* Columbia: Columbia University Press.

Roller, L. H., & Waverman, L. (2001). Telecommunications Infrastructure and Economic Development: A Simultaneous Approach. *American Economic Review, 91*(4), 909–923.

Schumpeter, J. A. (1942). Capitalism, Socialism, and Democracy.

Sharkey, W. (1982). *The Theory of Natural Monopoly.* Cambridge: Cambridge University Press.

Shy, O. (2001). *The Economics of Network Industries.* Cambridge: Cambridge University Press.

Simon, H. (1955). A Behavioural Model of Rational Choice. *Quarterly Journal of Economics, 69*(1), 99 ss.

Spulber, D., & Yoo, C. (2009). *Networks in Telecommunications: Economics and Law.* Cambridge: Cambridge University Press.

Stigler, G. (1971). The Theory of Economic Regulation. *Bell Journal of Economics and Management Science, 2*, 3.

Stiglitz, J. (2008). Forward. In G. Roland (Ed.), *Privatisation: Successes and Failures* (pp. ix–xx). New York: Columbia University Press.

von Hayek, F. (1978). Competition as a Discovery Procedure. In F. A. von Hayek (Ed.), *New Studies in Philosophy, Politics, Economics and the History of Ideas* (pp. 179–190). Chicago: Chicago University Press.
Weatherill, S. (2013). *EU Consumer Law and Policy*. Cheltenham: Edward Elgar Publishing.
Zamir, E., & Teichman, D. (2017). *Behavioral Law and Economics*. Oxford: Oxford University Press.
OECD. (2016), Being an Independent Regulator, The Governance of Regulators, OECD Publishing.
OECD. (2017b) Creating a Culture of Independence: Practical Guidance against Undue Influence

PART I

Pro-competitive and Pro-consumer Rules

CHAPTER 2

The Architecture of the European Regulatory Framework

Abstract European institutions have complemented the liberalisation of telecom markets with a harmonised regulatory framework hinging on pro-competitive independent regulation. The new EU regulatory approach has converged on tools and principles of competition law, which have been progressively internalised in this framework. In order to counterbalance the market power of the incumbent, asymmetric obligations requiring minimal behavioural restriction must be imposed. In addition, harmonised rules and consistent application across the Internal Market have been pursued by means of various supranational, horizontal and vertical coordination mechanisms. The regulatory framework is enforced at the national level by independent regulatory authorities. Independence from market players and national governments have been key to minimising regulatory failures and providing additional support to an EU enforcement coordination and harmonisation of rules.

Keywords New regulatory approach • Proportionality principle • Internalisation of competition principle • Asymmetric regulation • Regulators' independence • Legal harmonisation

A. Manganelli, A. Nicita, *The Governance of Telecom Markets*,
Palgrave Studies in Institutions, Economics and Law,
https://doi.org/10.1007/978-3-030-58160-2_2

2.1 The New Regulatory Approach

During the liberalisation process, a new regulatory approach was developed in Europe, where pro-competitive regulatory measures—i.e., *substantial liberalisation*—played a crucial role in governing the transition to effective competition. The new regulatory approach was adopted and progressively refined through a few directives under the standard harmonisation procedure (art. 114 TFEU) to establish the EU internal market.[1] In 2002, a few directives[2] outlined the so-called New Regulatory Framework (NRF)—a new regulatory approach whose principles are still in force, after the amendments in 2009[3] and a final overall revision in 2018 under the European Electronic Communications Code (EECC).[4]

The NRF explicitly defined competition promotion as an objective of the regulatory activity, jointly with protection of consumer interest and the consolidation of the internal market, in order to provide a substantial feature to liberalisation.[5] This involved an *internalisation of the competition principle*, as an *endogenous constraint* to regulation, whereas within the formal liberalisation directives, based on art. 106 TFEU, the competition principle worked as a mere *exogenous constraint* to (intrusive) public rules, as those granting special and exclusive rights (see Sects. 1.2 and 1.3).

The NRF is a harmonised regulation system, pursuing a competitive outcome in the liberalised market and comprising: (a) entry regulation, introducing a general authorisation regime, and (b) economic access regulation, mandating access obligations to incumbent networks in order to support competitive third-party entry into the market. The NRF also defined a complex, but effective, (c) institutional setting, based on a decentralised enforcers, i.e., national independent regulators (NRAs),

[1] The harmonisation under art. 114 TFEU requires a legislation procedure involving the Commission proposal and the final approval by the Council and the Parliament and may create new obligations for Member States.

[2] The main acts of the NRF were the Framework Directive (FD) 2002/21, establishing a harmonised framework; and the specific directives—Access Directive, 2002/19; Authorization Directive, 2002/20; Universal Service Directive, 2002/22; Privacy Directive, 2002/58.

[3] The 2009 amendment package included the Better Regulation Directive: 2009/140; Citizens' Rights Directive and BEREC Regulation: 2009/1211.

[4] Directive 2018/1972, to be considered along with the new BEREC Regulation 2018/1971

[5] Art 8(2,3,4) FD repealed by art 3(2) EECC.

integrated in a coordination system working at both vertical (with the Commission) and horizontal (with all other NRAs) levels.[6]

The NRF also covers (d) social regulation, such as consumer protection, digital privacy and universality of service. For the latter, the NRF focuses on *universal service obligations* (USO) to be imposed on undertakings in order to grant consumer access to those (baseline) services that are considered socially indispensable, even if not commercially profitable for the providers. This provision is to protect citizen interests by counterbalancing the social effect of liberalisation where "the liberalisation of the telecommunications sector and increasing competition and choice for communications services go hand in hand with parallel action to create a harmonised regulatory framework which secures the delivery of universal service."[7]

Traditionally, universal service concerned the fixed voice services and, since 2010, basic functional internet access. After, there followed an intense debate at EU and national levels about also including broadband connectivity. The *Broadband for all* objective was initially enshrined in soft law, such as investment targets (Sect. 6.1). Then, finally, the EECC explicitly states that "the concept of universal service should evolve to reflect advances in technology, market developments and changes in user demand", thus including the provision of affordable broadband internet access for all consumers. The EECC did not set an EU-wide minimum speed because this could have excessively distorted market dynamics, leaving to national level the specification of what kinds of services, quality levels and objectives of the supply to include within the universal service.[8]

As for the definition of the USO, NRAs are entrusted with task of targeting the company(s) required to provide universal service obligations. If the net cost of service obligations incurred by the designated companies is deemed excessively high, NRAs can introduce a compensation mechanism in favour of that provider (either through public funds or by distributing this cost over all the providers of electronic communications services).[9] Furthermore, the EECC allows NRA intervention in order to make internet retail prices accessible for low income and vulnerable consumers. This

[6] De Streel (2008).
[7] Recital 210 EECC.
[8] Art 84 EECC.
[9] Art 89 and 90 EECC.

occurs through direct financing of consumers or by requiring some operators to adopt special tariffs.[10]

Looking at access regulation under the economic theory of *property rights*,[11] an access obligation on an incumbent restricts the bundle of (lawful) uses or entitlements he has on its network infrastructure and increases the entitlements of access-seekers: moving from a property rule to a liability rule. *Property rules* give the owner of a good the power to prevent any other individual from interfering in the exercising of their rights, thus transfers of entitlements must take place with the consent of the owner. *Liability rules* allow third parties to secure entitlements on the good, subject to a compensation (usually set by the authority), regardless of the owner's consent.[12] As a matter of fact, regulation re-defines the uses and entitlements that are at the basis of opposing claims (i.e. *bilateral externalities*): on the one hand, there is the incumbent's interest to exclusively use its property; on the other, there are the access-seekers who want to use the incumbent's network at fair economic conditions in order to sustainably compete in the market. In the regulatory assessments and decisions, access-seeker protection is justified by the social cost of monopoly.

The whole array of potential uses of the property rights involved in the regulation is not completely known by regulators, also due to asymmetric information. Moreover, the uses change over time because of technological evolution, market structure dynamics, consumer preference shifts, and so on. It is, in fact, rivalry of evolving uses and claims which are considered by regulators in defining and periodically re-defining the balance between different private interests and public interest.

The above dynamic process is embraced by the European telecom regulatory framework which includes rules (a) establishing endogenous procedures for its periodic review,[13] with the intent of ex-post verification and correction of its effectiveness and efficiency, and (b) prescribing NRA periodic revisions of their market analysis and decisions,[14] in view of a co-evolution with businesses and markets dynamics.

A synthetic overview of the new regulatory approach, against the old one, is provided in the following Table 2.1.

[10] Art. 85 EECC.

[11] Demsetz (1967); Demsetz (1998).

[12] See Calabresi and Melamed (1972); Porrini and Ramello (2006).

[13] Art. 122 EECC.

[14] Art 67 (5) EECC.

Table 2.1 Old Approach vs New Regulatory Approach

	Old Approach	*New Regulatory Approach*
Objectives	Pursuing of public interest, not precisely defined, if not for the universal service aspects and the affordability of tariffs	Pursuing of public interest explicitly defined by the regulatory objectives: (a) competition promotion; (b) consolidation of internal market; (c) promotion of EU citizens' interests; and, since the 2018 EECC, (d) promotion of connectivity and take-up for very high capacity networks
Regulation Institutional Design	National central administration and/or regulatory functions within the legal public monopolist	National Regulatory Authorities (NRAs), independent from national governments, subject to vertical (with EC) and horizontal (with the other EU NRAs) coordination mechanisms
Entry Regulation	Legal monopoly: exclusive rights	No legal entry barriers, but a general authorisation (coupled with a licensing regime, as in the mobile industry, due to technical constraints, i.e. radio spectrum, and numbering resources)
Economic Regulation	Retail Price Control	Upstream access regulation, proportional to market power and only if competition law enforcement is not sufficient. Competition is assumed to discipline price, push quality and innovation. Incentive-based regulation aimed to push for productive efficiency.
Market structure	Single public operator	Multiple competing (private) operators
Social Regulation	Universal Service: Internal cross-subsidies from monopoly profits Consumer protection: subjection of user to the public monopoly (considered almost a public institution, with authoritative powers, rather than a market player). Limited national consumer protection legislation. Data protection and privacy: not considered a relevant policy issue.	Universal service: Competition with universal service obligations on one operator, if necessary, explicitly financed either by public subsidies or distributing costs over all operators. Consumer protection: extensive EU legislation progressively moving towards consumer empowerment Data Protection and Privacy: all communications must respect a high level of data protection and privacy, regardless of the technology used.

(*continued*)

Table 2.1 (continued)

	Old Approach	*New Regulatory Approach*
Regulation Evolution	Static approach: legal framework changed by exogenous and non-foreseeable interventions	Dynamic approach including endogenous procedures for periodic review of both regulatory acts and framework provisions.

2.2 Access Regulation: Pro-competitive and Proportional Intervention

As we have outlined, eliminating legal entry barriers does not automatically generate the conditions for alternative operators to compete effectively vis-à-vis incumbents. This is because a vertically integrated incumbent operator enjoys a competitive advantage based on the economics of telecom markets (i.e., economies of scale, scope, and network externalities), as well as on its ubiquitous network, built up over decades of monopolistic activity. This is difficult to be replicated by other operators in the short-medium term and is, therefore, considered an *essential facility* to operate in the market. Generally, a facility is considered as "essential" if it is socially less costly to provide the service on the existing asset than to invest in duplicating the asset.[15] As mentioned, this will occur in one of two cases: (a) the asset has certain natural monopoly characteristics, and/or (b) the investment cost of the asset has already been sunk and, therefore, the ex post avoidable cost of the input is much lower than the incremental cost of providing it.[16]

Consequently, downstream competition in the retail market for end-user services and the promotion of a level playing field are achieved by introducing ***asymmetric regulation***, i.e., special access (one-way) obligations, imposed only on the incumbent, aimed to counterbalance its market power and competitive advantage (Fig. 2.1).

[15] Here, we refer to the substantive situation described and not technically to the *essential facility doctrine*, which is a specific competition law approach adopted by US and EU courts to address refusal to share and interoperability issues. It belongs to the framework of refusal to deal and is based on the idea that a firm, which owns a facility practically impossible to be duplicated because of legal, technical or economic obstacles, has a duty to share its facilities with any market player, including competitors, seeking access (Oscar Bronner, Case C-7/97).

[16] Castaldo and Nicita (2007).

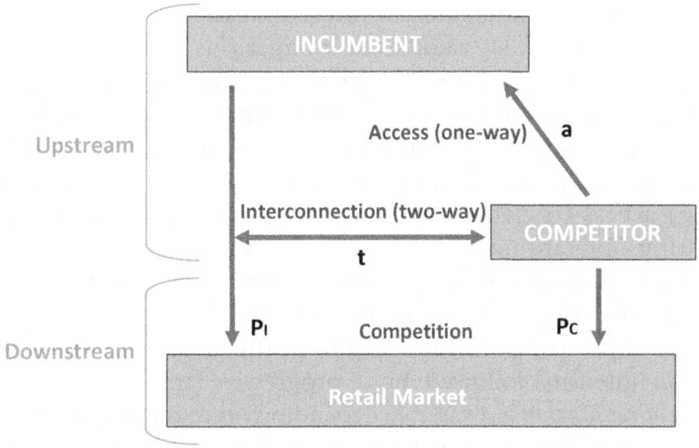

Fig. 2.1 Upstream and downstream markets

The internalisation of the competition principle in the NRF was implemented also by making regulation converge on competition law principles, case-law and tools and making both of them work as complements.

First, a regulatory action can be initiated only when general competition law is deemed to be insufficient to address the underlying market problem. Moreover, regulatory ex-ante obligations can be imposed, after a market analysis procedure, only on operators having significant market power (SMP). This is a concept equivalent to a dominant position in antitrust analysis,[17] i.e. the position of economic strength that provides a firm with the ability to behave independently of its competitors, customers, suppliers and, ultimately, final consumers.[18] In addition, regulatory obligations must entail the minimal restriction of company behaviour necessary to remedy the market problem and to achieve the framework's objectives (***proportionality principle***).[19]

[17] Para. 12 of the European Commission (2018) communication on Guidelines on Market analysis and assessment of significant market power—C(2018) 2374 final.

[18] ECJ in Case 2/76 United Brands.

[19] Art 68(2) EECC "Where an undertaking is designated as having significant market power on a specific market as a result of a market analysis … national regulatory authorities shall, as appropriate, impose any of the obligations set out in Articles 69 to 74 … In accordance with the principle of proportionality, a national regulatory authority shall choose the least intrusive way of addressing the problems identified in the market analysis."

Finally, besides the fundamental interconnection and interoperability obligations (Sect. 5.3), regulatory obligations are similar to typical antitrust remedies for abuse of dominant position by an essential facility owner. In particular, the EECC gives NRAs powers to impose obligations concerning: (a) transparency, (b) non-discrimination, (c) accounting separation, (d) access to, and use of, specific network facilities, (e) access to civil engineering, and (f) price control and cost-accounting obligations.[20]

The implementation of the regulatory framework, under the principles of convergence with antitrust and proportionality, was expected to generate a wave of pro-competitive regulation, with an initial drastic increase of regulation intensity followed by a progressive attenuation (Fig. 2.2). *Intensity of regulation* is a variable depending on regulation scope, defined by those cases where regulatory intervention is needed, and how heavy or light regulator's hand is in those cases. Immediately after formal liberalisation (see also Figs. 1.1 and 1.2), regulation scope is extensive due to the pervasiveness of incumbent market power, touching upon all market segments, and heavy-handed in order to establish a level playing field.

Afterwards, some regulatory obligations can be progressively waived, and the residual measures can become increasingly light-handed, reducing the intensity of regulation, as healthy market dynamics and empowered consumers become gradually able to discipline the market. Ultimately,

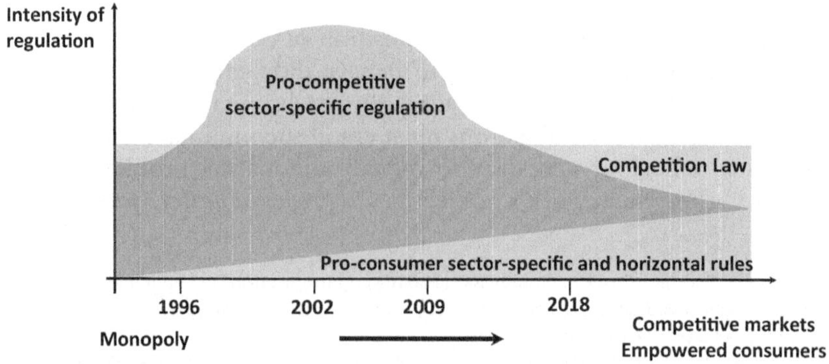

Fig. 2.2 The wave of pro-competitive sector-specific regulation

[20] Art. 69—74 EECC.

anti-competitive conducts become much less likely because of the market pro-competitive evolution, and can be in any event effectively addressed with the sole antitrust enforcement.

Consistent with this vision and expectations, in 2003 almost all upstream and downstream retail market segments were identified by the European Commission as susceptible of market regulation and were actually regulated in almost all European countries. Later, the number of market segments that required such market reviews has progressively fallen from 18 to 4. The other market segments, including all downstream retail markets, have since been declared effectively competitive. Hence, currently, with some national variance, ex-ante regulation has been limited to a few key wholesale upstream infrastructure bottlenecks, such as fixed telecom access and call termination (see Sect. 5.3).[21]

2.3 Vertical Separation as a Last-Resort Pro-competitive Remedy

An additional regulatory remedy for incumbent significant market power is functional separation, along with a voluntary separation commitment procedure.[22] These remedies were enshrined in the 2009 revision package, which introduction was a first shift from the original NRF concept. In the 2002 framework, access regulation was in fact deemed sufficient to (a) produce a pro-competitive market outcome and, as described in Sect. 5.2, to (b) progressively lead new entrants to invest in alternative infrastructures which would finally result in a facility-based competition with the incumbent. This shift took place because, notwithstanding ex-ante regulation, some vertically integrated incumbents managed to implement non-price discrimination practices. A non-price practice is a discrimination about the quality of the upstream input provided to competitors compared to what the incumbent provides to its own downstream divisions. Those practices are possible and relevant because alternative operators heavily rely on access to network of the incumbent—the exact extent depends on the competitor investment and consequently what kind of access services buy from incumbent (e.g., resale, bitstream and local loop unbundling—LLU). Such practices have been for example smaller or inconvenient space for competitors' equipment within incumbent's sites,

[21] Cave et al. (2019).

[22] Respectively art 77 and art. 78 EECC.

delay in network reparation when competitors' customers are involved, failures or delays to transfer a customer on the competitor's network, and so on. Non-price discrimination practices proved to be highly detrimental to competition, because they are much more difficult to detect and prevent by ex-ante regulators than price practices.

Therefore, vertical separation has been introduced as an additional possible remedy to ensure a non-discriminatory access to all downstream operators including the vertically integrated operator's own downstream divisions. After 2009, regulators are empowered to impose vertical separation, but only "in exceptional cases" where there has been "persistent failure to achieve effective non-discrimination in several of the markets concerned, and where there is little or no prospect of infrastructure competition within a reasonable time-frame after recourse to one or more remedies previously considered to be appropriate."[23] Vertical separation is therefore considered as a *last resort remedy* because (a) it is a very intrusive structural remedy whose application must be well delimited and restricted according to the proportionality principle; and (b) it involves a very difficult economic trade-off between allocative efficiency and productive efficiency.

Regarding the latter, as mentioned, vertical separation supports pro-competitive outcomes, in particular, by reducing the incentive for anti-competitive (price and non-price) discriminatory behaviour and the regulator's asymmetry of information, thus, making it easier to detect any discrimination practices. However, on the other hand, vertical separation would impact incumbent productive efficiency since a vertically integrated company (a) has no duplication of fixed costs, (b) *making* its access network can reduce ex-ante transaction costs involved in *buying* access services in the market,[24] and (c) can take full advantage of its scale and scope economies. Moreover, vertical separation could alter the incentives for a company (especially in a monopolistic situation) to invest in and upgrade its network. Finally, vertical integration is a solution for the management of incomplete contracts and ex-post transaction costs in the presence of asset specific investment, thus avoiding the *hold-up* problem, i.e., ex-post appropriation of return by a vertical trading partner.[25]

[23] Recital 202 EECC.

[24] The efficient decision of a firm to "buy" in the market or to autonomously "make" defines the boundaries between a vertical integration and vertical separation. See Coase (1937).

[25] Williamson (1985); Nicita and Pagano (2005).

These are the reasons why mandatory vertical separation has never been implemented in Europe. Nevertheless, there have been cases of voluntary separation resulting from more or less explicit "moral suasion" backed by the threat of the application of the non-voluntary vertical separation remedy or a competition law intervention. Just as an example, this was the case in the UK, where in 2017 a voluntary legal separation was agreed between British Telecom and Ofcom, after having Ofcom announced in 2016 to open a functional mandatory separation procedure.[26] Up to now, in EU there have been six cases of voluntary separation under to regulatory framework provision (i.e., in Czech Republic, Denmark, Ireland, Italy, Poland and United Kingdom), whereas, as said above, no functional separation obligations have been imposed.

There are different possible levels of separation, e.g., functional, legal and proprietary separation,[27] associated to different methods *of* equivalence of access (i.e., *equivalence of outputs* vs *equivalence of inputs*).[28] Consequently, private-ordering experiences which have taken place so far by committing to a voluntary separation have ranged in scope, intensity and approach, trying different ways to disentangle the regulatory trade-off. In some cases, relief from other regulatory measures have been traded against separation commitments. As we will see in Sect. 6.2, this is the same principle underlying the code provision about wholesale-only operators.

[26] In the UK, the national regulatory authority, Ofcom, has continuously focused on non-price discrimination practices implemented by British Telecom. Ofcom imposed a functional separation remedy in 2005, creating an upstream wholesale division—*Openreach*. In 2017, a deeper separation took place, with Openreach governed by an independent board and run by independent management still remaining in the BT group. For the UK case see Cave et al. (2006).

[27] Cave (2006).

[28] Equivalence of outputs means that the products the SMP operator offers to alternative operators and its own retail business are comparable in terms of functionality and price, although different systems and processes may be used. The Equivalence of Inputs model is more stringent and require applying exactly the same prices and processes to upstream wholesale products for competitors and for incumbent retail businesses.

2.4 Harmonised Regulation Toward a Single Internal Market

In addition to the promotion of competition, another key objective of the new regulatory approach is to strengthen and consolidate the European internal market, also by harmonising national legislations. A harmonised regulatory framework and its consistent application across the Internal Market has been a very important policy objective for the European telecom industry. This was deemed essential to enable trans-European networks and services to flourish. Against this backdrop, the high opportunity "cost of non-Europe"[29] has progressively been considered unbearable by all EU institutions.

Nevertheless, the enforcement of the EU telecom rules was mainly maintained decentralised. Currently the main enforcers are independent national regulatory authorities (NRAs), which are called on to carry out market analysis and impose regulatory obligations to SMP companies active in their national jurisdictions. In addition to national political reasons,[30] under an economic perspective, a decentralised enforcement provides benefits in terms of the effective use of dispersed information, as a national enforcer may have easier access to and a better understanding of the relevant information from national undertakings than a centralised enforcer. Moreover, a decentralised enforcement system may favour dynamic efficiencies because it entails a learning process that, through the comparison of the costs and benefits of different regulatory practices, leads to the emergence of the most efficient rules and best practices.[31]

The choice of the optimal degree of enforcement decentralisation involves important trade-offs. In order to find an efficient balance, a hybrid system has been progressively defined within the NRF, reaping most decentralisation benefits, mitigated by some centralised actions and effective coordination mechanisms. Indeed, some positive features of centralisation have been obtained through coordination (either vertical or

[29] See EU Parliament Research Service (2019).

[30] The European Commission in all reform procedures (2002, 2009 and 2018) has always proposed a EU centralised enforcer, an EU regulator. These proposals always encountered the opposition of (some) Member States, wishing to maintain regulatory powers at a national level, which led to always maintaining a decentralised setting, though mitigated by increasing coordination mechanisms.

[31] This has been highlighted by the literature on regulatory competition. see, for instance, Sun and Pelkmans (1995); Ogus (1999).

horizontal), while a negative effect of coordination is the potential impact on procedures complexity and timing.[32] Ex-ante centralised coordination and harmonisation mechanisms, as well as an ex-post coordination involving the Commission and NRAs, were established to minimise an inconsistent application of EU law.[33]

Ex-ante vertical coordination mechanisms are enacted through centralised actions by the European Commission. Firstly, a periodic *Recommendation on Relevant Product and Service Markets*[34] defines what are (in principle) the markets susceptible to ex-ante regulation by NRAs. This is a preliminary act of the regulatory process, where the NRAs define the actual relevant markets by possibly adding different product markets to the recommendation list. Any deviation from the list of relevant markets defined by the Commission must be justified, passing the so-called *three-criteria test*: "a market may be considered to justify the imposition of regulatory obligations … if all of the following criteria are met:

(a) high and non-transitory structural, legal or regulatory barriers to entry are present;
(b) there is a market structure which does not tend towards effective competition within the relevant time horizon, having regard to the state of infrastructure-based competition and other sources of competition behind the barriers to entry;
(c) competition law alone is insufficient to adequately address the identified market failure(s)."[35]

Furthermore, the European Commission can adopt harmonisation recommendations or decisions in the case of consistent and persistent divergence in national implementation.[36] Binding harmonisation decisions have always been very limited, while some harmonisation recommendations,

[32] A further positive effect of coordination is to help mitigate risk of regulatory capture and undue influences, therefore coordination supports independency, as it is true also the opposite relationship: independency supports supranational coordination (see next Sect. 2.5).

[33] See Manganelli et al. (2015).

[34] Under art 64 EECC. The first recommendation (2003/497/CE) was substituted by Recommendation 2007/879/CE, and the latest one is Recommendation 2014/710/EU. In view of a new Recommendation to be adopted by 21 December 2020, in 2019 the Commission launched a public consultation.

[35] Art 67(1) EECC.

[36] Under art 38 EECC

as pointed out in Sect. 6.1, have played an important role in the evolution of a systemic regulatory approach (e.g. NGA Recommendation, Recommendation on Cost Accounting Methodologies, Termination Recommendation).

A consistent implementation of the regulatory framework is complemented by ex-post vertical coordination, setting the scene for an European Commission "control procedure" over NRA decision-making.[37] NRAs must notify to the Commission their draft decisions and, in case the of "serious doubts" on their compatibility with the general framework, the Commission can open a so-called *phase II*—involving the NRA, the Commission and BEREC (The Body of European Regulators for Electronic Communications)— and potentially end up with a veto, blocking the NRA decision-making process.[38]

BEREC was established in 2009 under the framework revision[39] and is a network of EU regulators working as a horizontal coordination mechanism. BEREC is made up of all NRAs and, besides its role in the "phase II" procedure, it focuses on developing and disseminating regulatory best practices among NRAs. With this aim BEREC issues reports based on NRAs experiences, and provides advice to the Commission, the European Parliament, and the Council. BEREC has gained an increasing responsibility in complementing the framework provisions with regulatory guidelines. Albeit representing soft-law, BEREC guidelines have prominent importance, especially when the framework directive explicitly demands BEREC to make some general provisions operable by specifying regulatory principles or procedures expressed in general terms.

An additional positive aspect of centralisation/coordination is the internalisation of *institutional externalities.* It is quite possible that decentralised enforcement generates externalities on other decentralised enforcers. This is the case, for example, for international roaming, where a single decentralised enforcement activity involves a positive externality on other NRA(s). International roaming are upstream and downstream mobile services provided when a national end-user is abroad, within the EU, and his MNO has

[37] Under art. 32 and 33 EECC,

[38] The Commission veto is limited to NRAs' market definition and SMP identification. The 2009 review extended the Commission intervention also to regulatory obligations, by allowing it within a complex procedure to issue a recommendation. The EECC gives the possibility for the Commission to veto also measures related to symmetric access obligations (art 61.3 EECC) and co-investments commitments (art. 76.2 EECC).

[39] Regulation 1211/2009 repealed by Regulation 2018/1971

to interconnect with a foreign MNO in order to originate and terminate calls. National regulatory intervention at the wholesale level would produce benefits for foreign firms, whereas, limited exclusively at the retail level would create margin squeeze issues. As a consequence, within this framework, positive externalities entail a sub-optimal level of enforcement action (leading to under-restriction and free-riding).[40] This is exactly what happened among NRAs in EU, where each single NRA did not have any incentive to unilaterally regulate the international roaming prices imposed by domestic companies on foreign companies. Consequently, (most) national regulators did not intervene at all. This situation represented a clear failure of the existing coordination mechanisms and the overall framework, which had to be tackled by a centralised act of the EU institutions, which stepped in with the roaming regulation. This EU regulation has increasingly lowered the roaming charges, finally abolishing retail roaming surcharges (Roam Like At Home policy) in June 2017.[41]

2.5 Independence and Accountability of National Regulatory Authorities

In the overall framework design, supranational (horizontal and vertical) coordination is supported by the independence of each NRA from the respective national government. The effectiveness of the European coordination mechanisms can be considered as inversely correlated to the possibility of governments to give directions or instructions to regulators. For example, an institution like BEREC and its functions would not be imaginable if it would be made up of government ministerial departments. Regulator independence is therefore synergic with the objectives of legal harmonisation and the creation of the Single Market.

As anticipated, the Telecom Regulatory Framework has progressively imposed on Member States stronger regulatory independence requirements. Regulators are required to be independent from a) market players, being "legally distinct from and functionally independent of any natural or

[40] Deporter and Parisi (2006).

[41] In order to prevent misuses, this regulatory measure has been subject to a "fair use policy", i.e., capping the data roaming and possibly allowing some mobile operators may to continue applying a small roaming surcharge, if NRAs think the RLAH approach is not sustainable for them. Regulation (EU) 531/2012, as amended by Regulation (EU) 2120/2015.

legal person providing electronic communications networks, equipment or services"; and b) any other public body, including national governments. For the latter, NRAs "shall not seek or take instructions from any other body" and "only appeal bodies set up in accordance with Article 31 EECC shall have the power to suspend or overturn decisions by the national regulatory authorities".[42] Moreover, board members of NRAs "may be dismissed during their term only if they no longer fulfil the conditions required for the performance of their duties which are laid down in national law before their appointment".[43]

Regulator independence (and technical expertise) is a crucial aspect of the overall liberalisation outcome as it screens regulation from possible undue interference and/or opposition from national governments. This is crucial, in terms of a competitive level playing field, especially when there has been a liberalisation without privatisation (Sect. 1.5), i.e., national governments retain ownership or control of companies (usually the former monopolist) competing in the market with private undertakings. In this case, the framework specifically mandates Member States to "ensure effective structural separation of the regulatory function from activities associated with ownership or control".[44]

From an economic angle, rationales for independent regulation are significant and essentially lie in the goal of addressing market failures while trying to minimise the risks of some institutional *regulatory failures*, in the meaning of a public intervention that does not ultimately produce a socially desirable outcome in the market. In particular, the establishment of an independent authority allows (depending on the actual implementation) to:

(a) benefit from specific technical expertise and greater flexibility and speed of decision-making compared to traditional ministerial administrations;[45]
(b) decrease the risk of time-inconsistent regulation as they enjoy a longer mandate than political bodies (parliaments and governments) and their decision-making is not subject to the short-term preferences of political and electoral cycles (long-term policy commitments);[46]

[42] Art 3.2 and 8.1 EECC. Nevertheless, "this shall not prevent supervision in accordance with national constitutional law."

[43] Art 7(2) EECC

[44] Art 6(1) EECC

[45] See Thatcher and Stone Sweet (2002).

[46] Levy and Spiller (1996); Shepsle (1991); Armstrong and Sappington (2007); OECD (2016).

(c) ensure a credible regulatory environment, thus giving the correct incentives for investments and other long-term strategic choices by companies[47] and increasing the general compliance with rules;[48]
(d) create a level of government closer to sector-specific market players and users, so to better collect relevant information dispersed in the market, by also allowing supply and demand-side stakeholders to participate in the regulator's decisions-making via public consultations on draft regulatory acts.

Public consultations are mandatory,[49] in this respect, in order to give interested parties the opportunity to comment on the draft measures and address the interests of citizens. Consultations are to collect information and acquire a comprehensive understanding of the different views in the market (consumers included). Moreover, they provide enhanced transparency on the decision, as comments must be taken into account, possibly addressing them in a consultation report. Ultimately, considering the independence from top-down political influence, stakeholder participation in decision-making somehow results in re-creating a bottom-up link with society, companies and citizen interests.

In light of this, participation is also an accountability tool, which supports the other accountability provisions included in the framework, as the obligation to issue an annual public report of the activity, and in some Member States to also directly report to the Parliament.[50] In short, independence and its positive effects cannot justify a public power to be unaccountable or irresponsible. Beside being subject to the fundamental right of appeal against thier decisions (usually before administrative courts), independent authorities are accountable to stakeholders participating in their decision-making process, and, in a general way, to the public opinion and political institutions.

References

Armstrong, M., & Sappington, D. (2007). Recent Developments in the Theory of Regulation. In M. Armstrong & R. Porter (Eds.), *Handbook of Industrial Organization* (Vol. III, pp. 1557–1700). Amsterdam: Elsevier Science Publishers.

[47] Dixit (1996); Cambini and Rondi (2010).
[48] Majone (2001).
[49] Art. 23 and 24 EECC
[50] Art 8(1) EECC.

assessment. Telecommunications Policy, 32(11), 722–734.
Calabresi, G., & Melamed, A. D. (1972). Property Rules, Liability Rules, and Inalienability: One View of the Cathedral. *Harvard Law Review, 85*(6), 1089.
Cambini, C., & Rondi, L. (2010). Incentive Regulation and Investment: Evidence from European Energy Utilities. *Journal of Regulatory Economics, 38*, 1–26.
Castaldo, A., & Nicita, A. (2007). Essential Facility Access in Europe: Building a Test for Antitrust Policy. *Review of Law and Economics, 3*, 83–110.
Cave, M. (2006). Six Degrees of Separation: Operational Separation as a Remedy in European Telecommunications Regulation. *Communications & Strategies, 64*, 89.
Cave, M., Correa, L., & Crocioni, P. (2006). Regulating for Non-Price Discrimination the Case of UK Fixed Telecoms. *Competition and Regulation in Network Industries, 1*(3), 391–415.
Cave, M., Genakos, C., & Valletti, T. (2019). The European Framework for Regulating Telecommunications: A 25-year Appraisal. *Review of Industrial Organization, 55*, 47–62.
Coase, R. (1937). The Nature of the Firm. *Economica, 4*(16), 386–405.
De Streel, A. (2008). Current and Future European Regulation of Electronic Communications: A Critical.
Demsetz, H. (1967). Toward a Theory of Property Rights. *American Economic Review, 57*(2), 347–359.
Demsetz, H. (1998). Property Rights, in the New Palgrave Dictionary of Economics and the Law, 144-55.
Deporter, B., & Parisi, F. (2006). The Modernization of European Antitrust Enforcement: The Economics of Regulatory Competition, George Mason Law Review, 309-13.
Dixit, A. K. (1996). The Making of Economic Policy. A Transaction-Cost Politics Perspective.
EU Parliament Research Service. (2019). Europe's Two Trillion Euro Dividend: Mapping the Cost of Non-Europe, 2019-24.
Levy, B., & Spiller, P. (1996). A Framework for Resolving the Regulatory Problem, in Regulations, Institutions and Commitment.
Majone, G. (2001). Non-majoritarian Institutions and the Limits of Democratic Governance: A Political Transaction-Cost Approach. *Journal of Institutional and Theoretical Economics (JITE), 157*(1), 57–78.
Manganelli, A., Nicita, A., & Rossi, M. A. (2015). The Institutional Design of European Competition Policy. In E. Brousseau & J. M. Glachant (Eds.), *The Manufacturing of Markets: Legal, Political and Economic Dynamics*. Cambridge: Cambridge University Press.
Nicita, A., & Pagano, U. (2005). Incomplete Contracts and Institutions. In J. Backhaus (Ed.), *The Elgar Companion to Law and Economics* (2nd ed., pp. 145–161). Cheltenham: Edward Elgar.

OECD. (2016). Being an Independent Regulator, The Governance of Regulators, OECD Publishing.

Ogus, A. (1999). Competition Between National Legal Systems: A Contribution of Economic Analysis to Comparative Law. *International and Comparative Law Quarterly, 48*, 405.

Porrini, D., & Ramello, G. (Eds.). (2006). *Property Rights Dynamics: A Law and Economics Perspective*. London: Routledge.

Shepsle, K. A. (1991). Discretion, Institutions, and the Problem of Government Commitment, in Social Theory for a Changing Society.

Sun, J. M., & Pelkmans, J. (1995). Regulatory Competition in the Single Market. *Journal of Common Market Studies, 33*, 67.

Thatcher, M., & Stone Sweet, A. (2002). Theory and Practice of Delegation to Non-Majoritarian Institutions. *West European Politics, 25*(1), 1–22.

Williamson, O. (1985). *The Economic Institutions of Capitalism: Firms, Markets, Relational Contracting*. New York: Free Press.

CHAPTER 3

The Interplay Between Regulation and Competition Law Enforcement

Abstract Economic theory and public policy have generally embraced a clear-cut distinction between economic regulation and competition law enforcement as two separate public-interest policy tools, i.e., aiming at the maximisation of social welfare. Based on this approach, ineffective, incomplete, or unbalanced regulatory measures may lead to the need for an ex-post re-regulation and/or an antitrust intervention. How this relationship between regulation and competition law enforcement is designed and implemented (as complements in the EU or as substitutes in the US) has an important impact on the evolution of competitive dynamics and market outcome. Moreover, the interplay between sector-specific regulation and competition law enforcement is just one side of the competition policy spectrum, which is, overall, far more complex. It is comprised of additional crucial dimensions, i.e., decentralisation vs centralisation and judicial vs administrative enforcement, which further exacerbate the complexity of the institutional design and influence the overall pro-competitive market outcome.

Keywords Regulation • Competition law • Institutional design • Multilevel enforcement • Judicial enforcement

A. Manganelli, A. Nicita, *The Governance of Telecom Markets*, Palgrave Studies in Institutions, Economics and Law, https://doi.org/10.1007/978-3-030-58160-2_3

3.1 The Institutional "Division of Duties"

Competition policy is traditionally envisaged as a comprehensive spectrum covering different structural and behavioural public policy measures. Usually, the aim of liberalisation and economic regulation is to build a competitive market structure while competition law (antitrust law in US) and state aid rules work to preserve it.[1] Competition law prohibits anti-competitive conducts such as (i) agreements between companies having as effect or as object to restrict or distort competition, (ii) abuse by a company of its dominant position in the market, and (iii) mergers implying a significant impedement to effective competition, particularly by creating or strengthening a dominant position.[2]

A fundamental issue behind the design of competition policies concerns the interplay between competition law enforcement and economic regulation. A few years ago, Dennis Carlton and Randal Picker wrote: "Antitrust can say no but struggles with saying yes…while regulators often have a hard time saying no."[3] This is an important piece of the mainstream narrative regarding the *division of duties*, that is the allocation of competences, between Antitrust and Regulation. This simple reading reflects the traditional well-known vision, where regulation positively defines what are legitimate company behaviours (*what to do*), whereas competition law negatively defines what are prohibited company behaviours (*what not to do*).[4] Regulation is adopted before a specific behaviour takes place (*ex-ante*) with the intent of shaping it, whereas competition law enforcement aims at inhibiting and sanctioning a certain behaviour that has already taken place (*ex-post*).

In economic terms, in *the public interest theory of regulation*, economic regulation and antitrust represent two different ways of maximising social welfare (long-term consumer welfare). "Regulation and antitrust aim at similar goals—i.e., low and economically efficient prices, innovation, and efficient production methods (…) Regulation, however, seeks to achieve them directly, whereas antitrust seeks to achieve them indirectly by promoting and preserving a process that tends to bring them about."[5]

[1] State-aid rules are also part of the competition policy aimed at preserving healthy competitive dynamics (see Box 6.1).

[2] Respectively under art 101 and 102 TFEU, and Regulation 139/2004, Merger Regulation, for the EU jurisdiction; and section 1 and 2 of the Sherman Act (1890), and provisions under Clayton Act (1914) and Federal Trade Commission Act (1914) for the US jurisdiction.

[3] Carlton and Picker (2007).

[4] See, for example, Posner (2001); Hovenkamp (1994); Wish and Bailey (2018); Geradin et al. (2012).

[5] Breyer (1982).

Table 3.1 Economic regulation vs competition law

	Where	*Why*	*Who*	*How*
Economic regulation	Necessary where market dynamics cannot produce socially desirable outcomes	*Public interest*: (a) solution to market failures (b) restore market ability to maximise social welfare	(a) National Regulatory Authorities (NRAs)	– Positive definition of legitimate behaviour, i.e., rules about *what to do* – Rules are enforced by NRAs before anti-competitive conduct in the market takes place (*ex-ante*)
Competition law	Necessary and sufficient where market dynamics can produce socially desirable outcomes	*Public interest*: (a) preserve market ability to maximise social welfare	(a) National Competition Authorities (NCAs) (b) European Commission (c) National Judges	– Negative definition of legitimate behaviour (prohibitions), i.e., rules about *what not to do* – Rules are enforced by NCAs after behaviour in the market takes place (*ex-post*)

Economic regulation is necessary (and desirable) only when market failures occur, when market dynamics cannot produce a socially desirable outcome. Conversely, antitrust supervision is necessary and sufficient where market dynamics are likely to produce these outcomes on their own.

A synthetic overview of the differences between economic regulation and competition law, according to few basic descriptive variable (i.e., where, why, who and how to regulate rather than relying on competition law) is provided in the following Table 3.1.

3.2 Policies Interdependence in Different Institutional Settings

The above ideal-type institutional separation has shown its limitations over the years as the boundaries of competition law enforcement and regulation practice have become more blurred. There are ongoing important changes about *why* to regulate, as the theoretical distinction between economic regulation and social regulation have begun to fade. This is exemplified by the competitive challenges arising with big digital platforms (see

Chap. 8). Moreover, as far as the traditional telecom sector is concerned, there have been evident overlaps and interdependencies related to *who*, *how* and *where* to regulate or enforce competition law.[6]

As for the modes of intervention (*how*), there are clear overlaps between regulation and antitrust when competition law enforcers encroach on the *ex-ante* timeframe or *what-to-do* type of prescriptions. This happens when competition authorities use remedial action or accept a stakeholder commitment as to their future behaviour.[7] Furthermore, *how*-type of overlaps take place whenever regulators care about governing what happens *ex-post* through disciplinary proceedings or by reconciling disputes. In the Electronic Communications sector, as explained in Chap. 2, the extent of overlaps has significantly increased due to the internalisation of the competition principle, implying an alignment of the objectives pursued by regulation and antitrust. It is worth recalling that a large part of NRA decision-making is grounded on concepts, categories and tools drawn from competition policy (such as the definition of relevant market, significant market power, and the regulatory remedies).

As anticipated, one of the most significant burning issues involving this matter is the "division of duties" and the boundaries between antitrust and regulation (*where*). The key question here is—Where economic regulation is necessary, is it also sufficient? Or, in other words—What is the scope of competition law enforcement where economic regulation is necessary? Different jurisdictions have adopted different institutional solutions to these questions. In particular, the current application of the US antitrust law and EU competition law provide diametrically opposite answers.

In the US, the solution is one of mutual exclusion, that is, regulation excludes competition policy intervention according to the legal principle *lex specialis derogat generali*.[8] This position was reached in the famous *Trinko* case in the United States, along with the Supreme

[6] De Streel (2007); Monti (2008).

[7] Council Regulation 1/2003 on the implementation of competition rules (REG). At art. 9 gives the Commission the power to accept commitments offered by market players and make them binding on them in order to bring an infringement to an end and to meet the concerns expressed to them by the Commission in its preliminary assessment. The commitments are formally offered by undertakings, yet are likely the result of more or less explicit "moral suasion" backed by the threat of the application of antitrust law.

[8] This a principle governing normative conflicts. It means that if a factual situation falls within the scope of two different pieces of legislation (placed at the same level in the normative hierarchy), then the law governing that specific situation or subject matter prevails on and overrides the law with a wider general scope.

Court,[9] according to which "even though an incumbent that fails to provide assistance to would-be competitors, under the obligations of the Telecommunications Act, may maintain its monopoly, it does not violate section 2 of the Sherman Act in doing so."[10] Differently, the EU legal ordering envisages the coexistence of competition law enforcement and sector-specific regulation.[11] This approach is enacted in the 2002 Regulatory Framework and, specifically, in the market analysis guidelines. In particular, "it cannot be excluded that parallel procedures under ex ante regulation and EU competition law may apply with respect to different types of competition problem(s) identified on the underlying retail market(s). In this respect, ex ante obligations imposed by NRAs on undertakings designated as having significant market power aim to remedy market failures identified and fulfil the specific objectives set out in the Framework. On the other hand, EU competition law instruments serve to address and remove concerns in relation to illegal agreements, concerted practices or unilateral abusive behaviour which restrict or distort competition in the relevant market."[12]

This approach has been confirmed more than once by EU case-law, for example in the seminal case regarding Deutsche Telekom, which has been crucial—along to other margin squeeze cases (see Box 3.1) to define the relationship between regulation and competition law. In the DT case, the German company was regulated by BENTZA, the German NRA, which had approved the wholesale charges and imposed a price cap on a retail basket of services. The European Court of Justice confirmed the Commission's ruling on a margin squeeze given that DT had a margin of discretion within regulation. Moreover, the ECJ refused to accept an antitrust conduct defence when the dominant firm enjoys some flexibility within regulation. In particular the ECJ said that "... the competition rules laid down by the EC Treaty supplement in that regard, by an ex post review, the legislative framework adopted by the Union legislature for ex ante regulation of the telecommunications markets."[13]

[9] US Supreme Court Decision of 13 January 2004, Verizon Communications Inc v. Law Offices of Curtis Trinko. Verizon was regulated and has been sentenced by the regulators for not having given access to its network.

In the decision the US Supreme Court refused to condemn Verizon for violation of section 2 of the Sherman Act and award private damages to Trinko. Moreover, the US Supreme Court refused to extend section 2 liability to regulatory violation.

[10] Blair and Piette (2005).

[11] Larouche (2009); Geradin and Sidak (2005).

[12] Para. 12 of the European Commission (2018) communication on Guidelines on Market analysis and assessment of significant market power—C(2018) 2374 final.

[13] Case C-280/08P 14 October 2010 par. 90.

Box 3.1 Margin squeeze cases in EU telecom markets

Margin squeeze can take place when a vertically integrated undertaking holding dominant position in the upstream market gives access to one or more crucial inputs to firms operating in a downstream market (refer to Fig. 2.1). In the EU telecom context, upstream access charges (a) are regulated (pro-competitive access regulation), but incumbents usually have discretion to set downstream prices (p_i) to which competitors must respond with a competitive downstream price (p_c) in order to gain customers in the retail downstream market. By leveraging the integration of the upstream and the downstream markets, the dominant firm can set a relationship between its upstream charges (a) and downstream prices (p_c) so as to "squeeze" the margin available for competitors ($\simeq p_c$-a-other costs) and ultimately drive them, also those equally efficient to the incumbent, out of the market. This practice may actually represent an abuse of dominant position. However, it has proved to be a complex task to define when a margin squeeze represents an abusive behaviour under competition law, especially because of the concomitant application of sector-specific regulation.[14]

In Deutsche Telekom (2010), the Court of Justice clarified that an abusive margin squeeze occurs when a dominant firm leaves an insufficient margin between the prices of its upstream and downstream products. Furthermore, the decision stated that the approval of wholesale prices by a national regulator did not exclude the duty to comply with competition law by raising downstream/retail price in order to take the abusive squeeze to an end, pursuant to Article 102 TFEU. Interestingly, this ruling contrasts with the approach adopted by the US Supreme Court in Trinko (2004) where Justice Scalia held that antitrust enforcement is of little help if there is already "a regulatory structure designed to deter and remedy anticompetitive harm". The divergence between EU and US law was further extended with the US decision in Pacific Bell (2009) stating that antitrust law does not forbid lawfully obtained monopolies from charging monopoly prices both at the wholesale and retail level.

(*continued*)

[14] Pisarkiewicz (2018).

Box 3.1 (continued)

In TeliaSonera (2011), the Court of Justice stressed that margin squeeze is a stand-alone abuse and should not be considered as a constructive refusal to supply under article 102 TFEU. Thus, there is no need of a duty to deal placed on the dominant firm for a finding of abusive margin squeeze. The Court pointed out that in order to assess the lawfulness of a pricing policy, price squeeze must have an anti-competitive effect which may potentially exclude competitors who are at least as efficient as the dominant undertaking. This means that the dominant firm should have been sufficiently efficient to offer its retail services to end-users, otherwise than at a loss, if it had first been obliged to pay its own wholesale prices for the intermediary services. Notably, the existence of a positive margin does not *ipso facto* exclude the finding of an abusive margin squeeze. In this respect, with the ruling in Telefonica (2014), the Court of Justice highlighted that an abuse is an objective concept. In order to establish a violation of art. 102 TFEU the pricing conduct at issue should have exclusionary effects on as-efficient competitors. The Court was satisfied that such an adverse effect was generated on the retail market for broadband web services.

In Slovak Telekom (2018), the Court of Justice dealt with two substantive issues involving margin squeeze. First, the Court stressed that, in the case of a positive margin, it was necessary for the Commission to show that any reduced profitability would make it more difficult for the operator(s) concerned to trade on the downstream market. Second, the Court upheld the use by the Commission of fully allocated costs ("FAC") rather than the usual Long-Run Average Incremental Cost ("LRAIC") to assess whether the pricing conduct squeezed the margin available to as-efficient competitors as there was no better data available (see Sect. 5.3 for the different pricing methodologies). Further, the Court confirmed that the "equally efficient operator" test must be based on hypothetical competitors and not on the dominant firm own assets (as well as the unused network capacity of the dominant undertaking) or perfectly efficient firms.

In Europe, sector-specific regulation is clearly not considered *lex specialis*, i.e., regulation does not prevail on competition law. On the contrary, according to the EU hierarchy of norms, competition law has priority as it is primary legislation (since it is enshrined in the EU Treaty), thereby—in case of contrast—it overrides sector-specific directives which are pieces of secondary law. This however does not mean that competition law is able to tackle market failures and deliver on its goals without the complementary action of ex-ante regulation.

3.3 The Law and Economics of Competition and Regulation Policies in Europe

From an economic perspective, the rationale of the Trinko case, descriptive of the US approach, is that regulation is assumed to completely solve market failures and fully prevent those private behaviours that competition law would deter and in case punish. In this respect, regulation and antitrust are considered perfect substitutes. However, this theory of the ***division of duties*** seems to be unsatisfactory because it loses sight of a fundamental point of the analysis, namely, the relationship between incomplete rules and the residual economic freedom left to undertakings by regulation.[15]

Any form of law and regulation is incomplete as it cannot foresee and regulate every possible "state of the world". This may well be an intended outcome aimed not to constrain excessively private behaviours and leave room for enforcement discretion and flexibility. Nevertheless, when it comes to regulated sectors, the positive economic freedom granted by regulation cannot be considered as scrutinised once for all. One also needs to be aware that regulation may not always completely address market failures, especially when innovative dynamics are in place, and asymmetric information is likely.

Under the European approach, regulation limits the positive freedom of regulated firms (by restricting the set of their possible licit behaviours) but can neither cancel it[16] nor discipline all the possible cases in

[15] Nicita et al. (2007); Pistor and Xu (2002).

[16] Otherwise violating the principle of proportionality and undermining the incentives to innovate and the benefits that would result for the society, as a whole.

which that freedom may result in anti-competitive effects. This is even more so when considering that regulators must produce a pro-competitive outcome by imposing the smallest restriction on an undertaking's behaviour (proportionality principle). Indeed, in the context of a proportionality assessment, regulators must balance and provide an efficient solution to the trade-off existing between—on one hand, the individual cost imposed on the firm by restricting its freedom and, on the other, the social cost of a potential abuse. This must be done by considering that the probability of the abuse is inversely proportional to the degree of restriction of the behaviour and directly proportional to the existing information asymmetries.

When exercising its *residual economic freedom* (given the existing regulation), a dominant firm is still subject to a *special responsibility* whereby any behaviour hindering market dynamics is forbidden.[17] Therefore, a regulated dominant company has to balance its residual economic freedom with its special responsibility in the market. Competition law enforcement in a regulated context applies, therefore, to that residual economic freedom, if resulting in an exertion of market power which negatively affects consumer welfare.[18] This interaction produces a re-allocation of the risk (cost) of the information asymmetry, from the social dimension to the dominant regulated company, entailing a specific further responsibility. This additional responsibility, based on the combination of antitrust and sectoral provisions, involves an obligation for the dominant operators to pro-actively cooperate (primarily by providing adequate and truthful information) with regulators in order to define efficient and effective pro-competitive rules. In this regard, it is very interesting that the EECC defines a new possible obligation related to timely notification by SMP undertakings to NRA about their plans to migrate from the legacy infrastructure to a new next generation network (see Sect. 6.1).[19]

Therefore, antitrust intervention in regulated markets does not represent a regulatory failure, but it is one side of a multi-sided institutional enforcement mechanism in which ex-ante regulation represents another.

[17] Case 322/81 Michelin v Commission [1983] ECR 3461.
[18] Amato (1997).
[19] Art. 81 EECC.

This is an institutional solution that tends to result in an efficient *re-definition of property rights*, where the *incompleteness* of the rules defining which authority is ultimately entrusted with the task is endogenous to the system, thus not creating disincentives ex-ante.[20] The economic freedom of a dominant firm is restricted vis-à-vis competitors and consumers in order to prevent an abuse of market power that in turn would harm competitors and consumers' economic freedom. Within this dynamic process of rivalry between incomplete rights, it is efficient that regulation and competition law enforcement function in a complementary fashion.

Of course, regulation could to some extent act as a substitute for antitrust, as advocated by the US Supreme Court, only in the event of completeness of the law or complete rights. Namely, in a scenario where the regulator enjoys full knowledge of all the possible uses that are connected to the exercise of the legal framework as well as of the actual conduct of undertakings in the market. It is easy to understand that this presumption is not realistic. Indeed, regulation is expected to be forward-looking. It looks to the future, but cannot predict, nor determine it.

Regulation and competition law enforcement have thus complementarity features: regulation, made up of incomplete rules, shapes dominant undertaking economic freedom, by restricting their set of permitted behaviour. By doing so, pro-competitive regulation increases the likelihood of the undertaking's compliance with competition law. In fact, without any ex ante constraints, the probability of abuse would be high(er). In short, this is the rationale justifying ex-ante regulation and the principles guiding the definition of markets susceptible of ex-ante regulation—the *three-criteria test*. Should then regulation turn out to be ineffective—not inducing the dominant company to respect its special responsibility—competition law intervention plays a complementary role, indicating what additional behaviour is not allowed and providing effective tools to sanction it.

In other words, regulation determines the scope of the economic freedom associated with incomplete rights on the basis of a given "state of the world". Competition law enforcement governs the special responsibility that comes from those states of the world that are not contemplated by *ex-ante* regulation. On its side, regulation can undergo an adaptive process

[20] See Nicita et al. (2007); Nicita and Pagano (2005).

(re-regulation), by considering the additional revealed knowledge. This means that complementarity between regulation and antitrust legislation shows dynamic features. Regulation defines the scope of the antitrust legislation (the scope of the special responsibility) and antitrust law defines the evolving field of regulation (the areas in which freedom may result in unjustified exertion of market power). So, the division of duties, the frontier between antitrust and regulation is a mobile one.

Undoubtedly, there is a risk that this co-evolution might turn into a cumulative process progressively imposing more and more restrictions on dominant companies' economic freedom. In order to minimise this risk, re-regulation must take due account of innovation and technological evolution, thereby continuously updating in light of competition law principles. It is here—about the definition of relevant markets and the role of innovation—that an open dialogue on the interplay of competition law and regulation should mainly take place, with a *rule of reason* approach. This is the main challenge regulators and policy- makers face, especially in markets where innovation is paramount, such as in digital ecosystems and the platform economy (see Chap. 8).

3.4 The Institutional Design of EU Competition Policy in Telecom Markets

The interplay between sector-specific regulation and competition law enforcement is just one side of the competition policy spectrum, which is, overall, far more complex. As well as the horizontal relationships between different sets of norms and enforcers, there are vertical relationships and coordination mechanisms. On the regulation side, as described in Sect. 2.4, these vertical coordination mechanisms work among NRAs, BEREC and the European Commission (Directorate General for Communications Networks, Content and Technology—DG CNECT). Similarly, on the competition side, they function among National Competition Authorities (NCAs), European Competition Network (ECN),[21] and European Commission (Directorate General for Competition—DG Comp).

The main difference in this twofold vertical dimension is that the Commission is the traditional enforcer of competition law, whereas, as

[21] ECN comprises all the EU competition administrative authorities (Commission and all NCAs) in order to "apply the Community competition rules in close cooperation".

described, sector-specific regulation enforcement is primarily decentralised. The competition policy system actually has a strong centralisation of competences and powers and many NCAs were established after the bulk of the Commission's antitrust decisions had already been taken, shaping the main lines of the antitrust approach at the European level.

However, policy developments have increasingly strengthened the vertical "multi-level" nature of the system.[22] In particular, the so-called *modernisation* of competition policy[23] has fostered the role played by NCAs. As a result, powers and responsibilities in relation to competition enforcement are currently shared among actors, interdependently placed at different governance levels. The multi-level system includes coordination mechanisms between NCAs and the Commission and provisions which guarantee strong interpretative consistency through a functional hierarchy. This is the case, for instance, for (a) the initiation by the Commission of proceedings for the adoption of a decision which relieves NCAs' concrete competences,[24] and (b) NCA rulings on competition cases already subject to a Commission decision, which cannot be overturned at the national level.[25]

Moreover, besides administrative bodies (the Commission and NCAs), national courts also apply competition law, following complaints by private parties. This is why judicial enforcement is generally called *private enforcement* to distinguish it from public enforcement run by administrative bodies. Competition law private enforcement adds another relevant horizontal dimension to the overall competition policy system. From a general standpoint, the interface between private (judicial) and public (administrative) enforcement is based on different objectives and tools which are generally considered complementary. While courts protect the "interests of private parties", NCAs and the Commission safeguard the public interest for a competitive process within the market. Thus, private and public enforcement are complementary in that the former ensures direct compensation to parties damaged by anti-competitive conducts. As competition private enforcement works on a case-by-case basis, it is not necessarily concerned with the general issue of preserving competitive market dynamics.

[22] Kerber (2003).

[23] The milestone has been the enactment of Regulation 1/2003 (REG).

[24] Art. 11(6) REG.

[25] Art. 16(2) REG.

However, public and private enforcement should also be considered substitutes from a twofold perspective. First, they are alternative tools for ensuring legal certainty by complementing sector-specific regulation. Second, judicial compensatory damages and administrative punitive sanctions are to some extent equivalent in terms of deterrence, and, thus, their actions should be coordinated in order to minimise risks of sub-optimal deterrence.[26] As a matter of fact, there are coordination mechanisms meant to reduce the possibility of inconsistencies and conflicts. This is the case for the leading role played by the Commission in the general enforcement of competition policy also regarding the judicial application of antitrust law. Courts must avoid giving decisions that would conflict with a decision already taken by the Commission, or even merely contemplated in ongoing proceedings.[27] Moreover, the European commission plays a counselling role (*amicus curie*) transmitting relevant information to national courts and giving (non-binding) opinions on the application of competition rules.[28]

Overall, in Europe, competition policy in the telecom sector is based on a dense web of vertical and horizontal relationships involving many actors, all concurring to define the institutional environment in which competition takes place. These relationships, governed by both formal and informal coordination rules, entail significant competence overlaps and strong interdependencies (Fig. 3.1).

Firstly, there is a horizontal coexistence of competition law enforcement (A+B+C) and sector-specific regulation (D), which are meant to work as complements. Secondly, within competition law enforcement, there is an additional horizontal dimension, i.e., public administrative enforcement (A+B) vis-a-vis private judicial enforcement (C), having both elements of complementarity and substitutability that requires vertical coordination mechanisms. Finally, there is a complex vertical dimension related to the (multi)level governance of public enforcement, both on competition law and on the sectoral regulation. The conflict between decentralisation and centralisation has been framed in different ways, essentially depending on the different genesis of the two sets of norms. As for the public enforcement of competition law, there is both a decentralised national body (A) and centralised EU body (B), with vertical

[26] See Manganelli et al. (2015).

[27] Art. 16(1) REG.

[28] Art 15 REG.

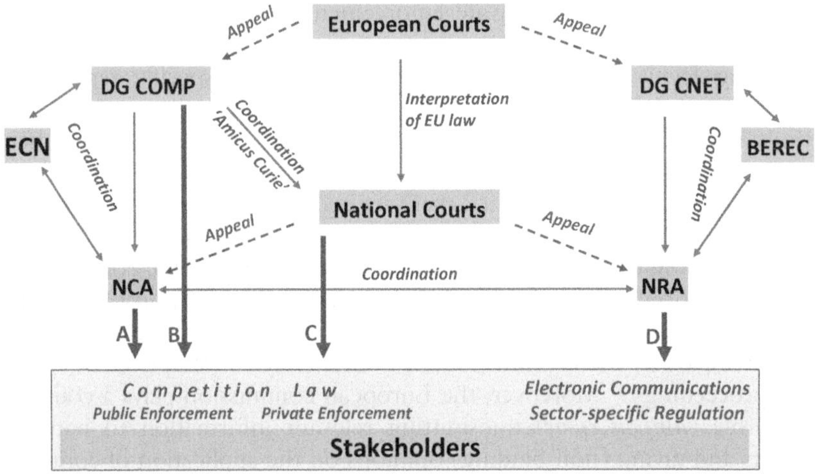

Fig. 3.1 The institutional design of European competition policy in the telecom sector

coordination mechanisms and the centralised enforcer playing a leading role. Electronic Communications rules are, instead, applied only at the decentralised national level (D), yet with strong vertical and horizontal supranational coordination mechanisms (Sect. 2.4). Of course, EU and national courts perform an essential role in judicial supervision by deciding on appeals of administrative decisions.

If one thinks that this dense web of relationships exists for each regulated sector, involving cross-sectoral interdependences and actors, it becomes blatantly clear that technological and economic evolution implying the convergence of previously distinct sectors (as described in Sect. 8.1 for media, telecom, and IT) may well require a rethinking of rules governing both markets and institutions.

References

Amato, G. (1997). *Antitrust and the Bounds of Power*. Hart Publishing.

Blair, R., & Piette, C. (2005). The Interface of Antitrust and Regulation: Trinko. *Antitrust Bulletin, 50*, 665.

Breyer, S. (1982). *Regulation and Its Reform* (p. 156). Harvard University Press.

Carlton, D., & Picker, R. (2007). Antitrust and Regulation, NBER Chapters. In *Economic Regulation and Its Reform: What Have We Learned?* (pp. 25–61). National Bureau of Economic Research.

De Streel, A. (2007). Antitrust and Sector-Specific Regulation in the European Union: The Case of Electronic Communications. In R. Dewenter & J. Haucap (Eds.), *Access Pricing: Theory and Practice* (p. 327). Emerald Group.

Geradin, D., & Sidak, G. J. (2005). European and American Approaches to Antitrust Remedies and the Institutional Design of Regulation in Telecommunications. In S. Majumdar, I. Vogelsang, & M. Cave (Eds.), *Handbook of Telecommunications Economics* (Vol. 2). North-Holland.

Geradin, D., Layne-Farrar, A., & Petit, N. (2012). *EU Competition Law and Economics*. Oxford University Press.

Hovenkamp, H. (1994). *Federal Antitrust Policy: The Law of Competition and Its Practice*. West Group Ed.

Kerber, W. (2003). An International Multi-Level System of Competition Laws: Federalism in Antitrust. In J. Drexl (Ed.), *The Future of Transnational Antitrust—From Comparative to Common Competition Law* (pp. 269–300). Staempfli Publishers.

Larouche, P. (2009). Contrasting Legal Solutions and the Comparability of EU and US Experiences. In F. Leveque & H. Shelanski (Eds.), *Antitrust and Regulation in EU and US* (p. 76). Edward Elgar.

Manganelli, A., Nicita, A., & Rossi, M. A. (2015). The Institutional Design of European Competition Policy. In E. Brousseau & J. M. Glachant (Eds.), *The Manufacturing of Markets: Legal, Political and Economic Dynamics*. Cambridge University Press.

Monti, G. (2008). Managing the Intersection of Utilities Regulation and EC Competition Law. *Competition Law Review, 4*(2), 123–145.

Nicita, N., & Pagano, U. (2005). Incomplete Contracts and Institutions. In *The Elgar Companion to Law and Economics* (p. 145). Edward Elgar.

Nicita, A., Rossi, M. A., & Rizzolli, M. (2007). *Towards a Theory of Incomplete Property Rights*. American Law & Economics Association Annual Meetings, n. 42.

Pisarkiewicz, A. R. (2018). *Margin Squeeze in the Electronic Communications Sector*. Kluwer.

Pistor, K., & Xu, C. (2002). Incomplete of Law. *New York University Journal of International Law and Politics, 35*, 931.

Posner, R. (2001). *Antitrust Law*. University of Chicago Press.

Wish, R., & Bailey, D. (2018). *Competition Law*. Oxford University Press.

CHAPTER 4

Evolution of Consumer Policy and Consumer Behaviour

Abstract Demand-side policies are crucial for complementing pro-competitive regulation (and competition law enforcement) in order to obtain a substantial liberalisation outcome. Therefore, policy actions should be aimed at empowering consumers as demand-side market players, e.g., by reducing switching costs and preventing lock-in phenomena. Those actions are integral to efficient market functioning, as pro-competitive regulatory measures, and not merely intended for the protection of vulnerable consumers. The European approach has always been to focus on consumer protection issues, both cross-sectoral and telecom-sector specific. Nevertheless, consumer empowerment policies may not be completely effective if actual consumer thinking, decision-making and behaviour—such as consumer inertia and other cognitive biases—are not duly targeted by regulation.

Keywords Consumer protection • Consumer empowerment • Switching costs • Consumer lock-in • Cognitive biases • Behavioural approach

A. Manganelli, A. Nicita, *The Governance of Telecom Markets*,
Palgrave Studies in Institutions, Economics and Law,
https://doi.org/10.1007/978-3-030-58160-2_4

4.1 Demand-Side Policy and Consumer Empowerment

In standard micro-economic theory, consumers play a pivotal role for well-functioning competitive markets. According to Adam Smith, "consumption is the only end and purpose of all production; and the interest of the producer ought to be attended to, only so far as it may be necessary for promoting that of the consumer."[1] The centrality of consumption in the ideal-type of the market in classical liberal economic theory derives from the idea that consumer free choice among alternative options in the market is the ultimate force that disciplines market power and inefficient allocation of resources. Thus, free choice among alternative products or services on the demand side in turn introduces the appropriate incentives for efficient decisions, by firms on the supply side, among alternative technologies and business methods.

This reasoning builds on the so-called *homo economicus* assumption: both consumers and producers behave fully rationally by maximising their own utility. They behave on the base of an optimal use of the available information and given the original endowments of economic resources. The institution of (perfectly competitive) markets generates optimal outcomes, under given conditions, precisely because rational agents meet and make choices, providing adequate information signals (on price, quantity, product availability, innovation and so on) to other agents in the market. By doing so they foster competition, providing businesses with strong incentives to deliver what consumers want.

Many OECD countries, and European countries in particular, have traditionally paid attention to consumer welfare issues, developing comprehensive general consumer protection rules, in order to shield the most vulnerable market players from exploitation by firms. On a more technical note (besides information disclosure obligations imposed to firms in order to compensate asymmetric information), those consumer protection measures result in a social regulation with a redistributive goal. However, the different role that consumers could play in influencing market dynamics has been almost neglected in a first period after market liberalisation. Indeed, the liberalisation wave of network industries in OECD countries have primarily focused on removing supply-side obstacles to develop competition in the market. However, for many years pro-competitive entry

[1] Smith (1776).

and access regulation was not completely supported and fuelled by freedom of choice of rational consumers, as stressed by the economic theory.[2]

In many circumstances, the freedom of choices enjoyed by consumers, even in a liberalised market, remained heavily constrained. Those limitations originate from monopolistic behaviours able to inhibit consumers to switch providers (*lock-in*) or to significantly increase their *switching costs.* These elements reflected into serious obstacles to consumers mobility and welfare, for markets where services are typically provided on a continuous basis and based on long-term contracts. Those impediments could have:

(a) contractual nature, like fixed long-term service contracts with very expensive anticipated exit clauses (lock-in);
(b) economic nature, like switching and exit costs concerning consumers sunk investments either in monetary (e.g., dedicated specific devices) or time-related terms (e.g., time spent to let other users know own phone number);
(c) a combination of both.

Progressively, policy makers and regulators started to dedicate more specific attention to the role of consumers for market functioning by developing so-called demand-side liberalisation policies. These policies were aimed to support competition dynamics also on the demand side, by empowering consumers to enhance the competitive pressure on businesses from the demand-side rather than merely create legal protection for individual rights. In their very essence, those policies aim at limiting consumer switching costs and contractual lock-in phenomena. They also try to enhance consumer trust and knowledge about products and services.

Current consumers policy is meant to address the unbalance in bargaining power between big firms and customers. Such policies build on the following two paramount objectives:

(a) to guarantee social fairness within the provision of services, by directly protecting consumers, especially vulnerable consumers;

[2] Armstrong (2008); Cseres (2008).

(b) to *empower* consumers by removing obstacles to the consumer free, rational, and informed choice, thus allowing them to play, especially those more sophisticated and active, that disciplining role in the market much praised by economic theory.

This distinction between consumer protection and empowerment is not always so clear-cut. Moreover, it entails a challenging *trade-off* that must be carefully addressed. Protecting consumers through excessive forms of paternalism would end up neglecting not only their rationality but also their maturity and even possibly incentivising moral hazard conducts from their side. More specifically, if public institutions protect consumers too much, they could get "lazy", renouncing to engage with markets, seek information and make the most out of the information and experience available. On the other side, empowerment cannot be effective without a certain level of protection; moreover, empowerment strategies definitely do not work for certain categories of consumers (e.g. vulnerable consumers), who need to be more openly and extensively protected as they are not able to engage on an equal footing with businesses, even under an extensive information disclosure regime.

A viable response to this policy trade-off is provided by the *liberal paternalism* approach proposed by Thaler and Sunstein in 2008.[3] Such concept may sound like an oxymoron, yet it is an attempt to balance consumer empowerment and protection by "*nudging*" them toward the most efficient choices. "A nudge is any aspect of the choice architecture that alters people's behaviour in a predictable way without forbidding any options or significantly changing their economic incentives. To count as a mere nudge, the intervention must be easy and cheap to avoid."[4] Those nudges build on information transparency and default options (*mixed default*) that work as standardised baseline rules to contractual relationships, taking advantage of consumers' inertia. Accordingly, the final decision made by consumers should be positively influenced by the nudge, not necessarily in the direction of the nudge but in the direction consumers prefer after the nudge has counterbalanced the external (or internal) constraints pushing him toward inefficient choices.

For example, what to do about unilateral changes of a consumer offer (i.e., combination of price and quantity) within a long-term contract? Is it more efficient to set an *opt-in* default (i.e., without consent, a provider

[3] Thaler and Sunstein (2008).

[4] As original defined by Thaler and Sunstein (2008).

cannot change the characteristic of the service) or an *opt-out* default (i.e., the user can be transferred to other service offer without prior consent, unless expressing an explicit refusal) along with effective information requirement? Or, further, is it better to set a dynamic default, indicating thresholds beyond which the option is triggered (*opt-in vs opt-out*)?

There are different possible levels and combinations of interventions. So, it is up to policy and regulatory reflection to understand what options, in different markets and areas, can effectively and efficiently combine protection and empowerment of consumers.

4.2 Consumer Policy in the Telecom Sector

Over the last years the European Union has paid great attention to consumer issues, and has developed a comprehensive consumer policy. Indeed, under the Treaty provisions, the Union is to contribute to the protection of consumer.[5] Furthermore, trustful, mobile, and empowered consumers are a condition for the substantial establishment of an EU-wide single market without any internal obstacles to the free movement of goods and services, which is at the heart of the European project.[6] Consumer policy was firstly based on traditional consumer protection and more recently also on consumer empowerment strategies. This process was led by the general consideration that "mature economies in which consumers are not empowered and where consumer detriment is high are more likely to remain on a low growth path. The focus of competition in these economies shifts from the fundamentals of quality and price to marketing strategies; competition weakens to the point where markets resemble monopolies and consumer confidence evaporates."[7]

The EU adopted both cross-sectoral (horizontal) consumer legislation and sector-specific one. The main cross-sectoral acts include the Directive on unfair contract terms, adopted in 1993,[8] the Directive on unfair commercial practice, adopted in 2005,[9] and Consumer rights directive,

[5] Under Article 169 TFEU. See for example Weatherill (2013).

[6] The protection of consumers is an essential part of safeguarding the EU internal market, which establishment and protection are set out as an objective in article 114 of the TFEU.

[7] See European Commission (2011).

[8] Directive 93/13/EC.

[9] Directive 2005/29/EC.

adopted in 2011,[10] as amended in 2019 by the Directive on better enforcement and modernisation of EU consumer protection.[11]

The horizontal directives have a twofold purpose:

(a) an effective achievement of market fairness, by protecting the weaker party in the commercial relationships;
(b) the establishment of the Internal Market through a harmonisation of national consumer protection rules.[12]

The Unfair Contract Terms Directive (UCTD) protects consumers against non-individually negotiated (standard) terms. Contrary to the requirements of *good faith*, these contractual provisions are likely to cause significant imbalance in the parties' rights and obligations to the detriment of the consumer.[13] This outcome can be due to asymmetries of information as well as lack of expertise and bargaining power in relation to the contract terms.

The Consumer Rights Directive (CRD) harmonises national consumer rules about certain aspects of contractual relationship (i.e., pre-contractual information requirements, formal requirements for the conclusion of contracts, detailed rules regarding the consumer's right of withdrawal), which are particularly relevant for the telecom sector with specific regards of distance and off-premises contracts.

The main objective of the Directive on unfair commercial practices (UCPD) is to boost consumer confidence, by enabling national enforcers to curb a broad range of unfair business practices taking place before, during and after a transaction. The Directive identifies: (a) a list of specific illicit practices;[14] (b) few general categories, i.e., misleading actions or omissions and aggressive practices;[15] and (c) a general principle under which is unfair any practice contrary to *professional diligence*.[16] A significant

[10] Directive 2011/83/EC.
[11] Directive 2019/2161/EU.
[12] Mateja and Micklitz (2017).
[13] Art 3 UCTD.
[14] Annex I UCPD.
[15] Art 6, 7, 8 UCPD.
[16] Art 5 UCPD. 'Professional diligence' means the standard of special skill and care which a trader may reasonably be expected to exercise towards consumers, commensurate with honest market practice and/or the general principle of good faith in the trader's field of activity, art 2 h UCPD.

example for prevention of non-contractual barriers to switching regards "use of harassment, coercion and undue influence."[17] This provision prevents traders from imposing any disproportionate non-contractual barriers detrimental to consumers who wish to exercise their rights under a contract (including the right to terminate the contract or switch to another product or trader).

These directives are horizontal in nature, covering all markets, and have allowed consumer policy to achieve a high level of harmonisation. They are principle-based so that provision can have a wide scope and are sufficiently broad to catch the evolution of services, sales methods, and companies. Nevertheless, they needed to be updated to address digital and technological developments. This revision was enacted in 2019 by the Directive on better enforcement and modernisation of consumer protection (see Sect. 8.4). This recent piece of legislation introduced a more effective enforcement, in particular involving sanctions (proportionality and increasing maximum fine) and remedies to which a consumer has access, such as monetary compensation, price reduction or replacement.

Next to those horizontal directives, covering all sectors, there is a detailed sector-specific legislation placing consumers into the specific context of telecom markets. Indeed, the NRF promotes citizens' interest by "ensuring a high and common level of protection for end-users through the necessary sector-specific rules and by addressing the needs, such as affordable prices, of specific social groups, in particular end-users with disabilities, elderly end-users and end-users with special social needs, and choice and equivalent access for end-users with disabilities."[18] The relevant telecom consumer policy act was universal service and users' rights directive,[19] which was part of the 2002 framework, then amended by the Citizens' rights directive[20] in the 2009 revision, and finally repealed in 2018 by the EECC.

As specified by the UCPD, in case of conflict between the provisions in the horizontal directives and rules regulating specific aspects of unfair commercial practices by means of specific sectorial legislation, the latter prevail in order to apply to specific sectoral aspects (*lex specialis* principle).[21]

[17] Article 9(d) UCPD.
[18] Art 3(2)d EECC.
[19] Directive 2002/22/EC.
[20] Directive 2009/136/EC.
[21] Recital 10 and Art 3(4) UCPD.

The EECC, beside revising previous sector-specific consumer protection rules to address evolution since the previous review in 2009, streamlined the sector-specific legislation in order to avoid overlaps in areas where horizontal rules alone can ensure an adequate level of protection for end-users. So now, telecoms end-user protection rules complement horizontal consumer protection law by addressing the specificities of the sector. The EECC thus focused on sector-specific consumer provisions by means of maximum harmonisation, meaning that member states may not impose more, or less, stringent provisions than those set out in the EECC.

Firstly, EECC enhanced transparency and contract information requirements for service providers. Companies are now asked to inform end-users in a clear, comprehensible, and transparent way about quality of the service, service conditions and prices in order to facilitate a consumers' well-informed choice and make it easy for them to compare between alternative offers and providers. This information must be supplemented by at least one independent digital comparison tool, which can be used free of charge in order to compare and evaluate different services in terms of price and quality of service performance.[22]

Specific provisions require that conditions and procedures for contract termination do not act as a disincentive for changing service provider. First EECC limits service contract duration to a maximum of 24 months. Moreover, if the contract provides for an automatic contract extension, consumers are entitled to terminate the contract at any time with a maximum one-month notice period, as determined by national law, and without incurring any costs except the charges for receiving the service during the notice period. Similarly, end-users can terminate their contract without costs upon notice of pejorative changes in the contractual conditions proposed by the provider.[23]

Finally, EECC regulate provider switching and number portability. In case of switching, providers must provide adequate information along the entire process and ensure the shortest transition time. In any event, the continuity of service must be ensured, unless technically unfeasible yet not exceeding one working day. End-users can retain or port a number to another provider any time for one month after the date of termination, without incurring in any charge. Porting of numbers and their subsequent activation must be executed within the shortest possible time, on the date

[22] Art 102 and 103 EECC.
[23] Art 105 EECC.

explicitly agreed with the end-user, and in any event not exceeding one working day from the date agreed.[24]

4.3 Switching Costs and Consumer Bounded Rationality

As a consequence of liberalisation and pro-competitive regulation, incumbents' market shares have quite constantly declined (see Fig. 1.3). However, dominant positions remain very difficult to challenge, especially in some countries. This is because only a portion of the potentially interested consumers tends to replace the usual supplier. This outcome can be the result of (a) the lack of expression of the consumers free choice once competition is introduced, or (b) perduring switching costs despite regulation, or (c) a deliberate choice not to move away from the historical provider for reasons related to service conditions or other consumers' preference.

As for the last point, it is worth highlighting that consumers mobility and growth of alternative operators signal a competitive market, whereas the opposite is not necessarily true. In other words, a market with empowered consumers can find a competitive equilibrium still having the incumbent operator a prominent share of the market. Indeed, what is primarily relevant for enhancement of competitive pressure in the market, from the demand-side, is potential switching, and not (only) the actual switching. Nevertheless, data about switching (especially time-series, cross-country or cross-sector comparisons) gives significative information about the competitive dynamics of a sector, especially in association with information about potential switching, which is however much more difficult to assess and precisely quantify.

The last issue of the *European Consumer Markets Scoreboard*[25] shows that a small percentage of telecom consumers switch suppliers, especially for the services offered via landline network. Next to the actual switching the scoreboard assess the possibility to switch provider easily, which represents a good indicator to measure the potential switching. Being able to easily switch from one provider to another allows consumers to optimise their choice and—by doing this—encourage competition in the market.

The Scoreboard highlights that churn and switch rates for fixed telephone services are lower than other communications services and other

[24] Art 106 EECC.

[25] European Commission (2018).

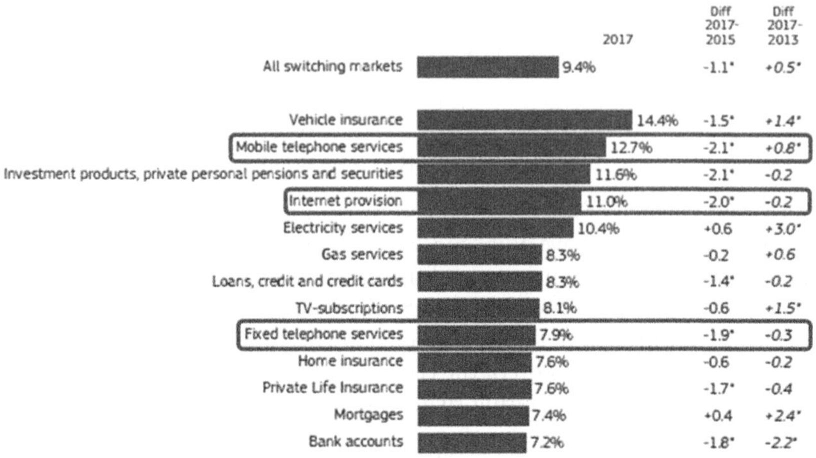

Fig. 4.1 Switching provider by market, 2017 (European Commission 2018)

sectors (EU average at 7.9%) and has decreased over time. Conversely, internet services and mobile telecommunications are doing much better, with 11% and 12.7% respectively (Fig. 4.1).

In addition, fixed telephone services have the lowest score, among industries with long-term subscription contracts, in the consumers' perception of an "easy switch" to another provider, described on the base of the question: "On a scale from 0 to 10, how difficult or easy do you think it was?" (Fig. 4.2). Of course this analysis describes consumers' perception, which may not well represent reality. However, this is indeed relevant, as it is the perceived cost of switching that affects the consumer's decision to whether actually switch or not.

Finally, considering the overall market performance indicator (MPI), which measures how well a given market performs according to consumers, all telecom markets score quite low.[26] Besides material problems with services and providers—all telecom markets, especially internet services,

[26] The Market Performance Indicator (MPI) is the central piece of the Consumer Markets Scoreboard. It is a composite indicator, which measures how well a given market performs according to consumers on the basis of: Comparability (i.e., how easy/difficult it is to compare offers), Trust (i.e., whether consumers trust that retailers/suppliers comply with consumer laws), Problems & detriment (i.e., what proportion of consumers encountered problems and extent of harm (including but not limited to financial loss), Expectations (i.e., whether a given market live up to consumers' expectations), and finally choice (i.e., whether consumers are satisfied with the number of retailers/suppliers on the market).

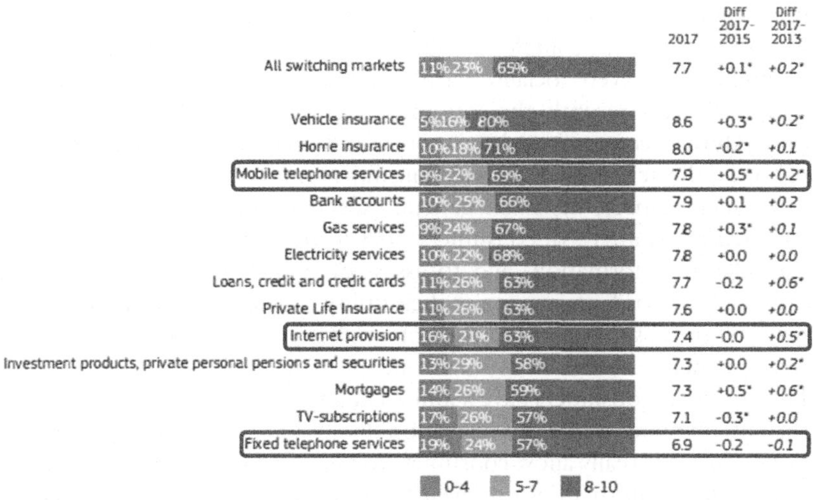

Fig. 4.2 Ease of switching provider by market, 2017 (European Commission 2018)

are the worst services markets among those considered, the main demand-side concerns involve a lack of trust by users (especially for mobile telephone services), which may hinder and slow consumers mobility.

Despite both European cross-sectoral consumer legislation and the telecom sector-specific have been working to empower consumers and remove demand-side obstacles, consumer inertia seems to be still a major hurdle for competition. Economists started to seek new answers by looking at the findings of cognitive and behavioural sciences.[27] The "common and fruitful simplification" that economic agents are perfectly rational has enabled economists to build powerful models to analyse a multitude of different economic issues and markets. Nevertheless, economists and psychologists have documented systematic deviations from the rational behaviour standards deployed by neoclassical economics.[28]

By adopting a cognitive and behavioural approach, the *rational choice* paradigm has been questioned. In fact, people are never perfectly informed and tend to oversimplify their decision-making process due to cognitive short-cuts and biases. On top of this, their actions are constrained by social

[27] As examples: Simon (1955); Thaler (1985); Akerlof and Schiller (2009).

[28] The Committee for the Nobel Prize in Economic Sciences (2017).

norms and inertia. Those cognitive "biases" cannot be neglected as otherwise even the most complete regulatory framework may end generating unintended or even socially harmful consequences. Somewhat surprisingly, only very recently the consumer policy has started to consider not only external (economic or legal) constraints on consumers, but also internal ones: the actual way in which consumers think.

In such respect, the main general aspect to consider is the so-called *bounded rationality*[29] of consumers, that is the inability to have access to the relevant information, and to assess even their own (intertemporal) preferences in order to maximise their utility. Consumers often do not factor in all the costs that a certain action implies, causing so-called *internalities* (namely, externalities that people impose on themselves).[30] Furthermore, consumers have a limited ability to process information, and this often implies that the information disclosed (even under legal obligations) does not really allow consumers to take a well-informed decision. In this context, information can even be too much (so-called "information overload"), and thus may be detrimental, confusing and not allowing consumers to clearly identify and select the relevant aspects. Also "too much transparency" when leading to information overload may increase transaction costs. In this sense "there is a clear trade-off between information completeness and simplicity, and also a potential crowding-out effect".

Finally, the econometric analysis and the behavioural and experimental economics demonstrate the existence of another factor, the so-called *status quo bias* or *inertia*. More specifically, even in the face of a possible improvement in the market, consumers would overestimate the benefits related to the existing position—or the magnitude of switching costs—and renounce to a positive opportunity. Consumers tend to stay where they are also due to informational and trust deficits. Often it is difficult for them (especially for certain categories) to understand the offers and/or to compare them, not to mention the risk of irrationally regretting choices made in the past by comparing them to heterogenous scenarios.[31]

This represents a huge challenge for policy makers. If consumers, due to inertia or other cognitive biases, are unable or unwilling to go for to the most convenient and innovative offers and supplier, competitive dynamics would be in any event negatively affected. In particular, competition in the

[29] Simon (1976).
[30] Allcott and Sunstein (2015).
[31] Motterlini (2014).

market is jeopardised and the liberalisation objective is partially frustrated as markets fails to discipline efficiently their players. In other words, the invisible hand of Adam Smith can do little if the visible one, that of the consumer, behaves randomly. Indeed, it is crucial for regulators and policy makers to understand consumers' behaviours and think what kind of regulation and consumer policy could tackle those issues. A mere increase of "traditional" pro-consumer regulation might not enhance competition but lead to additional market distortions.[32]

The most forward-looking regulators started to think that addressing traditional transaction costs is not enough. Indeed, it is essential to focus on the cognitive characteristics of market players, their attitude toward risk and uncertainty as well as their preferences about time management. For instance, consumers are likely to keep postponing tedious tasks such as switching operator or being distracted by contingencies.[33] Against this background, empowered consumers are only those who can effectively access relevant information and make the most out of them. This means being able to understand their own preferences and "reward" those businesses that best satisfy their needs.[34]

In order to address such challenges, in 2019, the OECD released a toolkit that provides policy makers and regulators with a step-by-step process for analysing a policy problem, building strategies, and developing behaviourally informed interventions.[35] Indeed, all regulators and policy makers are required today to update their approach towards consumer policy. A modern strategy should be able to combine traditional disclosure-regulation and cognitive-based regulation. By providing consumers with clear and transparent information tailored on their behavioural profile, regulators can ensure efficient market while preserving consumers' trust in firms and market functioning. Overall, consumers must be protected from abuses carried out by firms while enjoying the possibility to engage effectively with the market.

[32] Cafaggi and Nicita (2012).
[33] See OECD (2017).
[34] See UK Regulators Network (2015).
[35] OECD (2019).

References

Akerlof, G., & Schiller, R. (2009). *Animal Spirits. How Human Psychology Drives the Economy, and Why It Matters for Global Capitalism* (p. 21). Princeton: Princeton University Press.

Allcott, H., & Sunstein, C. (2015). Regulating Internalities. *Journal of Policy Analysis and Management, 34*(3), 698.

Armstrong, M. (2008). Interactions between Competition and Consumer Policy. *Competition Policy International, 4*(1), 97–147.

Cafaggi, F., & Nicita, A. (2012). The Evolution of Consumer Protection in the EU. In T. Eger & H.-B. Schafer (Eds.), *Research Handbook on the Economics of European Union Law*. Edward Elgar.

Cseres, K. (2008). What Has Competition Done for Consumers in Liberalised Markets? *Competition Law Review, 4*(2), 77–121.

European Commission. (2011) Staff working paper: Consumer empowerment in the EU - SEC (2011) 469 fin.

European Commission. (2018). *Consumer Markets Scoreboard*. European Commission.

Mateja, D., & Micklitz, H. W. (2017). *Internationalization of Consumer Law*. Springer.

Motterlini, M. (2014). *The Psyco-Economy of Charlie Brown*. BUR.

OECD. (2017). *Behavioural Insights and Public Policy: Lesson from around the World*. OECD.

OECD. (2019). Tools and Ethics for Applied Behavioural Insights: The BASIC Toolkit.

Simon, H. (1955). A Behavioural Model of Rational Choice. *Quarterly Journal of Economics, 69*(1), 99.

Simon, H. (1976). From Substantive to Procedural Rationality. In S. J. Latsis (Ed.), *Method and Appraisal in Economics* (pp. 129–148). Cambridge: Cambridge University Press.

Smith, A. (1776). *An Inquiry into the Nature and Causes of the Wealth of Nations*. London: Methuen & Co. (Book IV, Chapter 8, 49).

Thaler, R. (1985). Mental Accounting and Consumer Choice. *Marketing Science, 4*(3), 199–214.

Thaler, R., & Sunstein, C. (2008). *Nudge: Improving Decisions about Health, Wealth, and Happiness*. Yale University Press.

The Committee for the Nobel Prize in Economic Sciences. (2017). *Richard H. Thaler: Integrating Economics with Psychology*.

UK Regulators Network. (2015). *Consumer Engagement and Switching*.

Weatherill, S. (2013). *EU Consumer Law and Policy*. Edward Elgar.

PART II

Promotion of Investments and Technological Innovation

CHAPTER 5

Competition Enhancement and Investment Promotion

Abstract Telecom sector is a technological and investment-intensive industry. Telecom liberalisation policy has thus faced a difficult potential trade-off between competition enhancement and innovative investment promotion, simplified in the Schumpeter vs Arrow theoretical narrative. EU telecom directives frame this potential trade-off as an interplay between two possible types of competition: a competition based only on downstream services vs a competition of alternative infrastructures. This approach, known as the "Ladder of Investments", attempts to address the trade-off between long term investments and short-term competition. It aims at reconciling the positive short-term effect of service-based competition (static efficiency) with the positive long-term effect of infrastructure-based competition (dynamic efficiency). In this context, different models of access pricing and interconnection have been adopted as key regulatory tools to strike that balance by conveying and modulating different incentives and signals to the market.

Keywords Dynamic efficiency • Type of competition • Ladder of investments • Access pricing and interconnection

A. Manganelli, A. Nicita, *The Governance of Telecom Markets*,
Palgrave Studies in Institutions, Economics and Law,
https://doi.org/10.1007/978-3-030-58160-2_5

5.1 Competition and Investments: Economic Theory and Trade-offs

The telecom industry is a very capital-intensive and innovative technological sector. In the period 2016–2018, the main EU telecom and media companies invested, on an average, more than €7 billion per year, amounting to 27% of their revenues.[1] Therefore, investments, particularly in innovative infrastructures and services, are a crucial aspect of these markets and regulation could not underestimate this important aspect.

As outlined in Chap. 1 and Sect. 2.1, over the past thirty years under the OECD liberalisation wave, policymakers have designed a regulatory system to ensure downstream competition by means of pro-competitive access regulation (i.e., rules and obligations granting new entrants a fair and non-discriminatory access to incumbent facilities). Access regulation has been a keystone of the telecom market liberalisation in many OECD countries and in Europe, allowing for entry of alternative competitors and thus stimulating market competition. There is clear evidence that access-based competition has led to lower prices for consumers in the EU and higher penetration (see Figs. 1.3, 1.4 and 1.5); however, there have also been concerns that access obligations may have been detrimental to investments in new infrastructures, both for incumbents and new entrants. [2]

Indeed, a liberalisation where *full competition* is deemed to be a full-fledged superior alternative to any prevailing form of market power and competitive advantage is based on a static conceptualisation of efficiency, i.e. *allocative efficiency*. This is a scenario where sunk costs for building "the network" have been already sustained by the former legal monopolist, and competitors can enter the market just paying the cost of access to the network.[3] In a dynamic framework, where innovation and new network investments are relevant for social welfare, the overall effect of access-competition is more complex and uncertain. The economic approach to telecom industry has underlined "a general trade-off between promoting competition to increase social welfare once the infrastructure is in place and encouraging the incumbent to invest and maintain the infrastructure."[4]

[1] AGCOM (2019).
[2] See Cambini and Jiang (2009).
[3] Armstrong and Sappington (2006).
[4] Laffont and Tirole (2000).

Consequently, concerns about pro-competitive access regulation are based on the risks that it may negatively affect incumbents' incentives to invest, because both (a) an intensified competition would lower incumbent investment returns, and (b) the incumbent would bear the entire risk of the investment, while its potential benefits would also be enjoyed by access-seekers. Empirical studies showed evidence of an "inherent trade-off between access regulation and investment incentives", under which the intensity of access regulation is negatively correlated with incumbents' investments, while has a positive effect on aggregate new entrants investments (while negative effects on single entrants are possible depending on market circumstances).[5]

Along these lines, two opposing arguments on competition and investment promotion have taken centre stage in the regulatory debate following liberalisation, often oversimplifying the questions at stake. On the one hand, there is the pro-incumbent's vision supporting a light-handed regulation, if not a *regulatory holiday* or *forbearance*, leaving the incumbents with enough freedom of action to increase profitability and, consequently, its investments. On the other, there is the pro-entrant's vision advocating for more extensive regulatory restraints on incumbent market power and low cost-based access prices as the key to enable new investments in a competitive environment.

Both arguments are worth being considered and build on the lessons of Joseph A. Schumpeter.[6] and Kennet J. Arrow. [7] The former was of the opinion that a monopoly or a firm with significant market power has bigger incentives to innovate than firms facing competition in the market. This is essentially due to the monopolistic firm's greater ability to appropriate the benefits stemming from technological breakthroughs, by excluding potential competitors from enjoying those benefits. Moreover, larger firms have *deep pockets* (i.e., the financial capacity to make big investments in innovation). Therefore, Schumpeter's position underlies those positions envisioning unsolvable trade-offs between competition and innovation.

[5] Grajek and Roller (2012). This widely considered econometric work analyses 70 fixed-line operators from 20 EU Member States, in the period 1997-2006, finding access regulation to have a negative effect on total industry, i.e., \$16.4 billion, and individual carrier investment.

[6] Schumpeter (1942).

[7] Arrow (1962).

On the other hand, Arrow stressed how a monopolistic firm has less to gain from innovation because it faces opportunity costs that a firm in a competitive market does not bear—i.e., the risk of losing its existing pre-innovation monopoly profits (the so-called *replacement effect*). In short, an undisputed dominant firm does not have any strong incentive to innovate further as this is unlikely to improve its condition. Therefore, within Arrow's conceptual framework, it is firm competition, rather than monopolistic rents, that triggers innovation in a synergic way. However, Arrow also describes the *appropriability problem* that reduces incentives for companies to innovate in a perfectly competitive market. In economic terms, innovation is a *public good*[8] creating knowledge spill-overs. In other words, innovation generated by a company produces positive externalities across the market, which need to be internalised by public intervention in order to maintain incentives to innovate. This could be done either by means of public subsidies or, more commonly, through intellectual property rights (IPRs), entitling exclusive exploitation of innovation benefits (for a certain period), thus eliminating *free-riding* effects.

An in-between positioning can be found in a number of empirical works showing the existence of a non-linear relationship (*bell-shaped* or *inverted-U*) between competition and innovation.[9] According to these analyses, both low and high levels of competition would reduce the incentives to invest, and the point of maximum investment would be linked to an intermediate level of competition. Therefore, in order to promote innovation, regulation does not have to be linearly and indefinitely pro-competitive but, instead, should define and target an optimal intensity or *type of competition*.

In point of fact, the views of Schumpeter and Arrow are not necessarily conflicting.[10] Under an economic policy perspective, innovation has been widely considered a desirable outcome of competition which, in turn, was regarded as a process resulting in *dynamic efficiency*.[11] Schumpeter himself described the changing and cyclical relationship between competition and monopoly, where they represent the two sides of the same coin.

[8] Its consumption is non-excludable and non-rival and subject to free-riding problems, where consumers enjoy benefits of consumption without contributing to public goods production, resulting in a suboptimal production, and, finally, inhibiting the production when private costs exceed private benefits.

[9] Aghion et al. (2005); Briglauer et al. (2012).

[10] Kerber (2017); Baker (2007).

[11] Dynamic efficiency refers to the extent to which, over time, a firm introduces new products or new process of production, contraposed to static, i.e., allocative efficiency

Competition provides firms with the right incentives to innovate, while differentiating services and products in an effort to obtain monopolistic revenues.[12] Monopoly, on the other hand, attracts new competitors driven by the desire to erode away a part of those monopolistic rents. Hence, in Schumpeter's view, competitive dynamics are naturally devoted to establishing monopolistic positions and, vice-versa, monopolistic rents lay the foundation for new competition.

This vision of the relationship between innovation and competition has led to framing the tension between static and dynamic efficiency diachronically.[13] To deal with the trade-off, regulation should promote competition and enhance social welfare in the short term, while promoting innovation and investments in the long term. Accordingly, ways of conceiving regulation can vary over time, modulating the *optimal intensity* of competition rather than restricting the market entry of new actors. Indeed, there are two main ideal types of competition and a continuum between them. On the one hand, there is a *service-based competition* (SBC) (or intra-platform competition), where new competitors rely (only) on the incumbent's network in order to provide their services. On the other hand, there is a *facility-based competition* (FBC) (or inter-platform competition), where competitors build their own infrastructures and compete, as vertically integrated operators, *vis-à-vis* the incumbent.

Within the 2002 EU regulatory framework, there was a general leaning towards facility-based competition (also called infrastructure-based competition) mainly because of its strong incentives to spur innovation, both at the network and service levels, driving efficiency improvements.[14] However, facility-based competition was unlikely to take place in the short-term following liberalisation, due to the high investments needed to build new nation-wide networks. "Due to the high risk involved in investments with a high share of sunk costs, alternative operators are likely to follow a step-by-step approach, continuously expanding their customer base and infrastructure investments. The initial availability of the

[12] Schumpeter (1934).

[13] Curzon Price and Walker (2016).

[14] See Manenti et al. (2006). This study is based on 14 European countries and from period 2000-2004 find that competition in DSL market doesn't play significant role in broadband uptake, while inter-platform competition triggers broadband diffusion and lower unbundling prices have positive effect on broadband uptake. See also Bouckaert et al. (2010). This paper find assesses a sample of 19 OECD countries, finding that both inter-platform and intra-platform competition have positive effects on broadband deployment.

incumbent's infrastructure at low prices will make it easier for alternative operators to enter the market and develop a customer base. Equipped with a customer base, uncertainty is considerably reduced, and the operator may then be ready to take further investments."[15]

On this basis, policymakers tried to conceive and adopt a regulatory model able to reconcile those aspects, which took the name of *Ladder of Investments* (LoI) (see next Sect. 5.2).

The case of inter-platform competition between an incumbent and a cable operator is inherently different. Due to digitalisation and convergence (see Sect. 8.1), this kind of competition has become relevant in those countries with a legacy of analogue cable assets, where investment have been done for the digitalisation and the consequent possibility to provide broadband and telephone services (e.g., NL, BE). In other countries alternative cable infrastructures didn't develop at all. Overall, in 2018 in EU the coverage of digital cable operator was 45%.[16]

5.2 The Ladder of Investments: Temporal and Geographic Dimensions

The Ladder of Investment approach was the policy result stemming from the debate over the most appropriate way to address the trade-off between long-term investments and short-term competition.[17] It was intended to make service-based competition and facility-based competition work as complements in balancing competition and innovative investment. It reconciles the positive short-term effect of service-based competition (static efficiency) with the positive long-term effect of infrastructure-based competition (dynamic efficiency). In other words, the Lol approach aims to selectively and dynamically provide new entrants with the right incentives, with the right timing, on what type of competition to develop, whether it involves buying access services from the incumbent or making their own access services by investing in access network components.

This objective was implemented through a transitory "entry assistance" for new players, making it relatively easy for firms to access the incumbent's network and quickly develop an effective service-based competition

[15] European Regulators Group (2004).

[16] European Commission (2019).

[17] As for the original formulation of the approach see Cave and Vogelsang (2003); and Cave (2006).

and, eventually, reach a critical mass of the customer base (i.e., SBC as *stepping stone*). Once competitors are consolidated in the market, this entry assistance would progressively decrease, consequently, providing them with incremental disincentives to access incumbents' networks. This incentive mechanism is aimed at neutralising a *replacement effect* for the new entrants (similar to that described by Arrow for dominant firms), which would otherwise prevent competitors from investing and creating their own infrastructures in the medium/long run. This replacement effect is based on the opportunity cost for entrants to invest in new alternative infrastructures, which is created by an access regulation that allow entrants to *buy* access services (conveniently and permanently) rather than *make* them.

In realm of the Ladder of Investment theory, the main toolkits used by regulators to provide incentives for new entrants are:

(a) modulating access price, according to which prices are set at a relatively low level and then gradually increased, along with the development of the entrants' customer base;
(b) introducing a *sunset clause* approach, involving the complete removal of access regulation after a certain date.

An increasing access price or the sunset clause are supposed to provide regulatory incentives for new entrants to progressively move from Service Based Competition to Facility Based Competition, adopting different access options—the *rungs of the ladder*—marked by a progressively greater level of investments. Incentives are linked to each rung of the ladder, gradually reducing, for each company, the opportunity cost of leaving its current rung. It is then up to market dynamics to define what is the efficient level of investment and the sustainable level of competition in each rung—that is, within the inverted-U relationship between competition and investments.

A hierarchy of access options to the traditional incumbent's network, i.e., the ladder rungs, is shown below in Fig. 5.1. The access options range from (a) mere resale of services, requiring the least network investment, as operators receive and resell the wholesale input from the incumbent; through (b) bitstream access or other virtual access, to (c) Local Loop Unbundling (LLU)—in either a shared access form or a full unbundling, requiring competitors to invest in network infrastructures up to the local

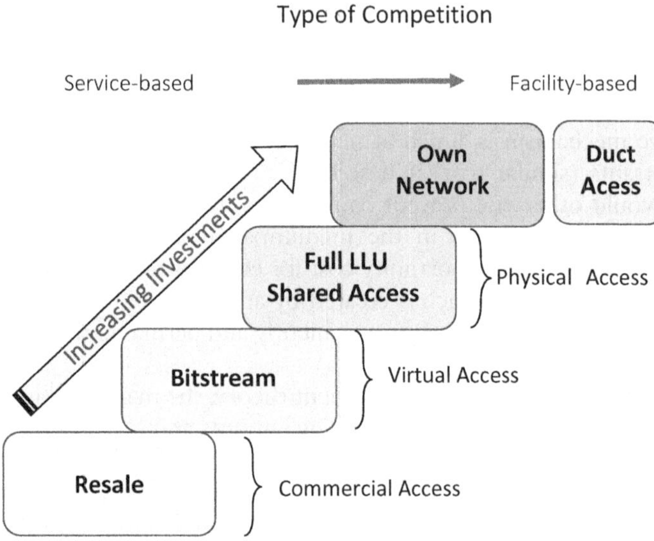

Fig. 5.1 The Ladder of Investment

exchange.[18] On the top of the ladder, competitors have developed (d) their own infrastructures to the end-user premises, and, thereby, no longer having to access the incumbent's network.

The LoI approach worked well under the first wave of liberalisation policy "as a means of implementing unbundling in a way which progressively promotes competitive providers' infrastructure investment in fixed networks".[19] The climbing of the ladder in the EU is described below by

[18] The local loop is the physical circuit connecting the end-user's premises to the local exchange, so it approximately coincides with the so-called *last mile* of the network. Unbundled access to the local loop was a regulatory milestone of the EU telecom industry, adopted before the NRF, under Regulation (EC) 2887/2000 and imposing on incumbent operators to meet reasonable requests from new entrants to access their local loop. While in full LLU setting the alternative operator access the entire local loop to offer voice and broadband services, in a shared access setting the entrant access solely the upper line's bandwidth to offer only broadband services. Bitstream is a virtual, not physical access, at a more elementary level of the value chain; the access point for bitstream is beyond the local loops and usually closer to the entrants' core exchange. See at Fig. 6.1, the legacy ADSL architecture.

[19] Cave (2014).

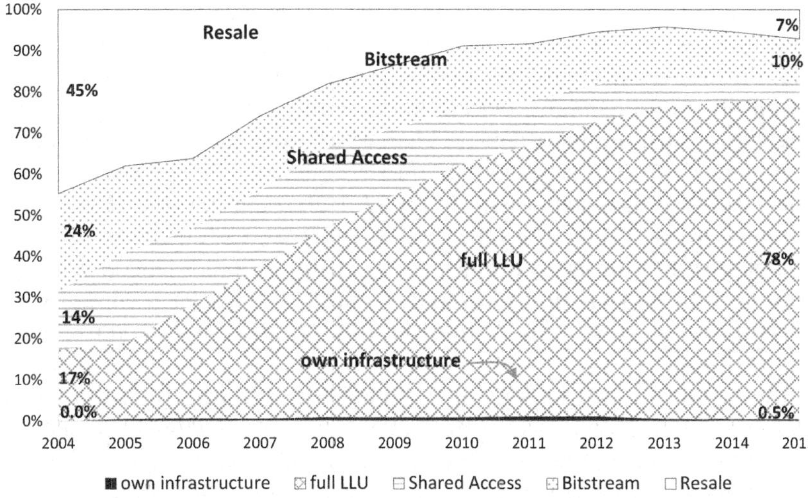

Fig. 5.2 Percentage of access type for new entrants in EU, 2004–2013

Fig. 5.2 showing the increasing percentage of LLU over time, and thus the change of competition type in the continuum between SBC and FBC.[20]

However, the LoI was later criticised from an empirical perspective as it was perceived to be effective only in the *short ladder*, pushing alternative competitors to the local loop unbundling (LLU) or sub-loop unbundling (SLU) rungs of the ladder, but not beyond. [21] Therefore, the critique was that, with few exceptions,[22] the LoI failed in making alternative operators develop their own infrastructure has not produced a market outcome where alternative operators developed their own infrastructure.

On a theoretical basis, the LoI has been criticised on three levels. Firstly, it has been argued that in order for the LoI approach to succeed, the regulators should always have access to the right information. Furthermore, they are required to have the skills needed for micro-managing the evolution of a market from service-based to facility-based competition.[23]

[20] Data from Communications Committee (2014).

[21] Bacache et al. (2014).

[22] There are few operators who have moved across the entire LoI; however, it is difficult to ascertain what part in this outcome depends on access regulation methodologies.

[23] Odale and Padilla (2004).

A second critique was based on the fact that service-based competition serves as a stepping stone to facility-based entry only if the replacement effect is neutralised, and if the regulator has the instruments to do so.[24]

Finally, even assuming perfect information and competences on the regulatory side, an important problem arises from the actual dynamic of competitor entry. In fact, in relation to a sequential entry of competitors, there is a time-consistency issue in the LOI. In order to preserve both the incentive to overcome the replacement effect for early entrants and the stepping stones for later entrants, access charges should depend both on time and the entry period so as to dynamically provide each entrant with the same access conditions.[25] However, this is quite complicated to obtain in the market.

As a point in fact, looking at the regulatory practice, regulators have favoured entry at each level of the ladder value chain (*splintering*) and for each type of technology, with neutral rules and incentives, according to the *technological neutrality* principle. Indeed, another founding principle of pro-competitive regulation in Europe was the so-called technological neutrality, according to which "national regulatory or other competent authority should neither impose nor discriminate in favour of the use of a particular type of technology" [26]

Moreover, regulators, with very few exceptions, tended not to adopt a clear-cut mechanism to neutralise the replacement effect, such as a sunset clause or an increasing access charge. On the contrary, access charges have often decreased over time (as shown in Fig. 5.3 below),[27] also due to the increasing cost-efficiency of incumbents enabled by incentive-based regulation.

Consequently, a diachronic climbing up the ladder's rungs was limited to "physiological" investments and market dynamics, following the initial

[24] Bourreau et al. (2010).

[25] Avenali, Matteucci G, Reverberi P (2010) Dynamic access pricing and investment in alternative infrastructures, International Journal of Industrial Organization Volume 28, Issue 2, March 2010, Pages 167-175

[26] Recital 25 EECC. The EECC declined the technological neutrality principle in a different way from the 2002 Directive, by specifying that it does not preclude taking into account that certain transmission media have physical characteristics and architectural features that can be superior in terms of quality of service, capacity, maintenance cost, energy efficiency, management flexibility, reliability, robustness and scalability, and, ultimately, performance, which can be reflected in actions taken with a view to pursuing the various regulatory objectives.

[27] Data from European Commission (2018).

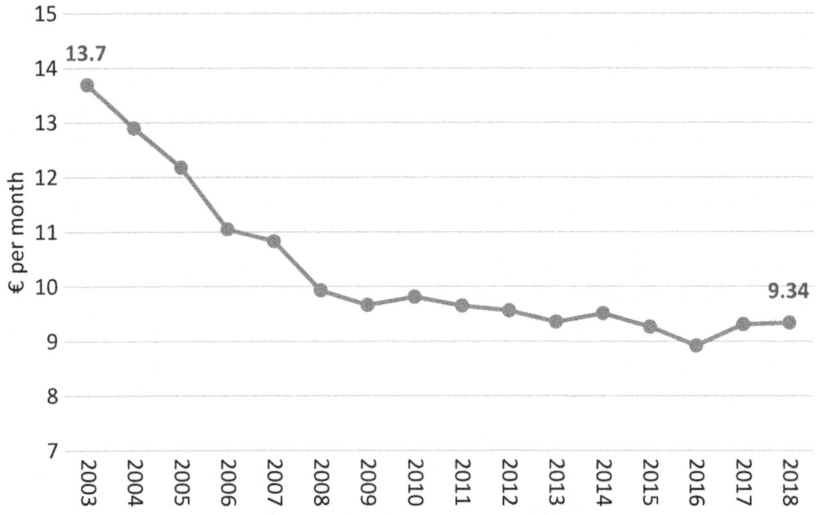

Fig. 5.3 Evolution of full LLU average price, 2003–2018

entry assistance. Conversely, no measure based on the LoI theoretical approach surfaces as a clear and widespread regulatory mechanism. However, a differentiation in investments and competition in the continuum between SBC and FBC has taken place, not on a temporal, but on spatial dimension, that is, at a geographical level.

Where liberalisation rules make it possible, competitive entry takes place only in profitable geographical areas and market segments, as for the classical *cherry picking* dynamic. In the EU telecom markets, there is not only a binary differentiation (competition vs non-competition) in the different geographical areas, but also a differentiation in the type of competition depending on the level of infrastructure development along the rungs of the investment ladder.

Market structure ranges between its extremes across areas, some areas with a monopoly infrastructure, with mere resale or no competition at all, while others have competing alternative end-to-end infrastructures. Notably, in this last category, very different *leader-follower* dynamics have taken place on access network investments where, in some cases, investments have been driven by new entrants, in others, investments have been induced by incumbents.[28]

According to this evolution, the European framework allowed NRAs to introduce both sub-national geographically relevant markets,[29] together with "differentiation in the appropriate remedies imposed in light of the differing intensity of competitive constraints" within a geographic market.[30] The underlying idea is that a geographically differentiated regulation may also have the potential to improve the trade-off between static efficiency (competition) and dynamic efficiency (investment incentives), and could do it more effectively and pragmatically than the LoI approach. Indeed, optimal regulation from a social perspective involves a differentiation in access prices according to local levels of infrastructure competition, as in a three-dimensional *ladder of investment* or *chessboard of investments*..[31]

5.3 Access Pricing as Key Element for Regulatory Balance

The telecom regulatory framework has adopted ex-ante obligations which reflect the ex-post remedies of the competition law toolbox to abuse of dominance (for instance, access obligation, transparency, non-discrimination, accounting separation and price regulation). Unlike antitrust enforcement, as previously highlighted, ex-ante regulatory obligations are forward-looking and must balance the different framework objectives, and not only pursuing competition promotion.

In this balancing, the key regulatory measure has been price regulation, given its flexibility and ability to transmit and modulate incentives and signals to the market. This ability enables regulators to pursue different objectives and (try to) address existing regulatory trade-offs. Economic

[28] Cave (2014).

[29] The definition of relevant product and geographic market, converging with competition law practice, has been embraced by the Regulatory framework since 2002. The actual definition of geographic subnational markets is however a more recent practice in most EU countries linked to the differentiation in investments and, thus, the intensity of competition in different areas.

[30] Recital 172 EECC

[31] Bourreau et al. (2015).

theory actually envisages multiple objectives for the regulatory definition of access charges, namely:

(a) allowing competition to emerge and encouraging an optimal level of competitive entry (efficient entry);
(b) providing incentives for incumbent cost-efficiency;
(c) encouraging an efficient utilisation of the incumbent network, and also encouraging efficient investments from the incumbent;
(d) providing the right signal to entrants about the *make or buy* alternative and promoting investments in their own network.

Access price works here as the key element to balancing competition enhancement and investment promotion. This balance, as we have seen in the previous Sect. 5.2, evolved from the LoI approach to the new European Electronic Communications Code (EECC), which promotes a more efficient remuneration of risk, co-investment incentives and, in specific situations where competition safeguards are in place, a more light-touch regulation.

Access price is also a key tool for fostering effective competition. To encourage efficient levels of market entry, access charges should not be too high, otherwise they would represent a competitive barrier. At the same time, they should not be too low, otherwise they would allow inefficient entry, ultimately resulting in an excessive and unsustainable downstream retail price competition. In turn, such an ecosystem would lead to inadequate remuneration of the incumbent infrastructure costs, and low investment incentives.

Using a simplifying assumption for perfect competition among new entrants in the retail market, an optimal access charge that maximises social welfare should be equal to marginal cost, $a = mc$.[32]

However, in a situation of high fixed costs, as in the telecom industry, an access charge equal to marginal costs is not sustainable for the incumbents, as lower than the average costs. In fact, regulating access to an essential facility should aim at allowing other companies to profitably enter the market, while, at the same time, allowing the incumbent to recover the fixed costs relating to the service provided.[33] Therefore, regulators are

[32] Marginal unit cost is the cost of one additional unit produced, i.e. the variation in total production costs that occurs when one additional unit is produced.

[33] Armstrong et al. (1996).

required to determine the level of the access price increase above marginal cost in order to compensate fixed costs and allow firms to break-even.

This approach is called *Ramsey pricing*, which is built with a double mark-up over the marginal costs (mc): $a = mc + mark\text{-}up_1 + mark\text{-}up_2$. Mark-ups depend on (i) the elasticity of demand of the entrant's retail service, and (ii) the substitutability of the services offered by the incumbent and the entrants. Ramsey prices would be a regulatory second-best option, but they have never been applied due to their inherent complexity and lack of information on price elasticity.

Regulators have had to follow a simpler implementation path, leading to *cost-plus* access charges, such as the *rate of return*, or *incentive-based regulation*, such as *price caps*.

In a *Rate of Return* (RoR) regulation, regulators define (and review periodically) the undertaking's *revenue requirement* and then set prices so that the revenue earned by the firm does not exceed the revenue requirement, therefore, restricting the amount of profit (the return) that the firm can earn. The revenue requirement is based on the firm's accounting costs calculated from total costs including all variable costs, plus a reasonable rate or return on the asset base. RoR regulation takes into consideration fixed costs to provide a fair remuneration for incumbents, while ensuring correct "make or buy" signals (provided competition downstream is intense). As RoR gives incentive to cross-subsidy, the remedy of accounting separation is necessary, as well as attempts to determine top-down costing associated with *fully distributed costs* (FDC). This means that all costs, including joint and common costs, are fully allocated (FAC) to all the operator's services/products according to a specified distribution/allocation criterion.

Admittedly, this methodology comes with two major drawbacks. First, RoR provides low incentives for cost minimisation (production efficiency), innovation and reduction of operating costs and, therefore, could lead to relatively high prices (compared to a competitive benchmark). Second, RoR increases the risk of incumbent bias for input consumption resulting in over-capitalisation, when the allowed rate of return on capital is higher than the cost of capital (*Averch-Johnson effect*[34]).

Access charge is also considered a fundamental tool to improve productive efficiency, by means of incentive-based regulation. Basically, regulators commit to allocating in favour of the essential facility owner (part of)

[34] Averch and Johson (1962).

additional profits coming from its cost-efficiency gains. In this situation, incumbents are incentives to reduce production costs and become more efficient.

Considering the above, the main alternative to ROR is *Price cap*. Price Cap methodology has been widely used in many countries, both for retail price and wholesale price regulation and adopted for the first time in telecoms in the UK in 1983.[35] Under this pricing mechanism, the regulator determines a cap, putting a limit on the firm's prices, but without limiting its profits. Price cap is also called RPI-X, as the cap is periodically adjusted depending on the Retail Price Index and a factor X, which usually reflects the productivity growth of the industry in relation to the rest of the economy for a specified period. So the cap on prices at a period p+1 is equal to $PC_{p+1} = PC_p$ (RPI-X). Over this time span, the average price of the specified basket of goods and services should not exceed the price cap, while the firm can adjust internal prices within the basket. Consequently, the firm has an incentive to reduce costs, and part of these cost reductions can be passed on to consumers via a periodic revision of factor X. The length of time between the reviews is of the utmost importance. While short lags imply that a greater fraction of surplus is transferred to consumers, but at lower incentives, long lags provide higher incentives for cost reduction, but postpone the moment when consumers can finally obtain their welfare share.

The drawback of price cap is that prices may diverge significantly from actual costs and may, thus, reduce allocative efficiency and welfare. Moreover, incumbents bear the risk of having a negative additional profit. In this case, there would be a switch of risk from consumers to companies and, therefore, a higher cost of capital. Finally, strong incentives to reduce costs may also imply incentives to reduce quality,[36] eventually requiring service quality regulation and heavier consumer empowerment rules. However, there is empirical evidence that Price Cap Regulation leads to more investment.

Another pricing model, widespread in the EU in the last years, is based on bottom-up *long-run incremental costs* (LRIC). Incremental cost is the additional cost that a firm incurs in providing a service or the cost that the company would avoid by deciding on not to provide the same service. The

[35] Littlechild (1983).

[36] Empirical evidence is mixed: with price cap regulation, quality seem to increase in some dimensions, while decreasing in others, relative to ROR. See Sappington and Weisman (2010).

Table 5.1 Access pricing methodologies: pros and cons

	Rate of Return	*Price Cap*	*LRIC / LRIC+*
Pro-competitive	Yes	Yes	Yes
Cost Efficiency	No	Yes	Yes
Allocative Efficiency	Yes	Yes/No	Yes
Dynamic Efficiency	No	Yes	Yes
Quality of Service	Yes	No/ Yes	No/Yes

evaluation of this cost is made from a long-term perspective where fixed costs are supposed to become variable and the company can choose the production factors efficiently, so as to minimise its cost functions. In particular, long-run incremental costs (LRIC) are a hybrid form of cost-plus regulation as a rate of return based on long-term incremental costs, assuming that the operator has become efficient. The advantages of the LRIC method are that it predicts volume and cost movement effectively and represents an economically rational approach to pricing cost-based services over time, while providing incentives for cost-efficiency. In order to consider a different extension of the perimeter of common and joint costs, a mark-up can be added to the pure-LRIC (*LRIC plus*).

A schematic overview of pros and cons of the access pricing methodologies described is offered at Table 5.1 above.

In addition to access regulation, the other fundamental upstream regulation is related to networks interconnection, which is also called *two-way access* (refer to Fig. 2.1). It involves a reciprocal access between two networks that rely on each other to deliver their services. *Networks interconnection* is a fundamental symmetric regulatory measure (applied to all operators regardless of their market power), as it is a systemic pre-condition for a liberalised market to function.[37] Network interconnection allows a customer of a network operator to communicate with customers of another network operator, and is a very different concept from unilateral access to the essential facility. More precisely, a network operator uses the network of another operator—the termination segment—in order to terminate the calls of its customers directed to the customers of the other network (*off-net calls*).

Interconnection internalises and maximises network effects (for users and society). Interconnection actually increases the number of active users

[37] Art 15(2a) and 60(1) EECC.

in the market, by allowing users from one network to interact with users from another competing interconnected network. Moreover, interconnection allows new entrants to compete against incumbents. Without this element, users would never subscribe to a start-up network operator, as this would prevent them from communicating with most telecom users. Incumbents could foreclose on new entrants, also by differentiating between on-net prices and off-net prices, thus creating new network externalities, tariff-mediated, and thereby partially vinifying the positive economic effects of interconnection. *Tariff-mediated externalities* would increase the costs faced by users for using a small network rather than a big one. As they would call the incumbent's customers, given its much larger customer base, they would end up paying a higher off-net price.[38]

In European countries, a *Calling Party Pays* (CPP) system, where only the caller pays, was adopted for the retail price of voice services. Accordingly, at a wholesale level, the caller's network pays the receiver's network to terminate the call. Here, incumbents could foreclose on new entrants increasing termination charges, i.e., *raising rivals' costs*. Consequently, *call termination* has been subject to price regulation, however, not only for the incumbents but for each operator. This is because the relevant market is identified as call termination on each single network. As each operator exclusively enjoys its own termination network, it is considered holding a dominant position on this market segment. Given the interconnection obligation and the incumbent's termination price control obligation, a small operator could also profitably raise their termination prices much above the competitive level. Due to this peculiar economic aspect, fixed and mobile termination markets were included in all EC recommendations on relevant markets and have been permanently regulated.

Call termination regulation has been marked by a twenty-year long burdensome regulatory process, and an incredible number of judicial disputes have taken place.[39] Similarly, divergent implementation approaches, and policy objectives surfaced, such as the financing of mobile markets and spectrum assignments through the very high fixed to mobile termination rates. Subsequently, in 2009, the Commission issued a harmonising Recommendation,[40] establishing common principles for remedying

[38] Hoernig (2007).

[39] All operators' termination was regulated, i.e., more than 1000 (!) dominant companies in Europe.

[40] Recommendation 2009/396/ EC on fixed and mobile termination prices in the EU

divergences in the regulation of fixed and mobile termination rates at national level. It was recommended that termination rates were based on the costs incurred by an efficient operator, implying that these tariffs would also be symmetrical, i.e., equal for all network operators. To this end, the Commission recommended the adoption of bottom-up models based on current costs and pure LRIC method. Finally, following the EECC new provision,[41] a Commission delegate directive will put an end to this *termination saga* by directly setting very low termination price caps.

The main problem with termination regulation was the assimilation of the two-way access to the one-way access, by means of an access pricing methodology similar to the that used to limit the high market power of the owner of an essential resource. The market failure in the termination bottleneck (of new entrants) does not originate from the essential nature of their termination facility, but from externalities. The users paying the termination on the retail market—the caller—do not choose the network where the termination will take place and, therefore, its price. In fact, the caller obviously selects the user to call, but only the latter chooses its own network. The indifference of the called party to a price that they indirectly select, but are not paying, weakens the normal economic forces operating on the demand side, which otherwise would oblige operators to compete also on termination prices. Moreover, the called party does not pay for the benefits they receive from the call—the *call externalities*. In general, if the caller does not internalise this externality—for example, in a closed group of users where each component takes into consideration the welfare of others in their decisions—an inefficient allocation of resources is generated. Namely, too few off-net calls are produced. Considering call externalities, the termination price that maximises the welfare is certainly below the marginal cost of termination, unlike a one-way access price.

Therefore, welfare maximising termination rate is $t = c_t - u_r$, where c_t is termination marginal cost and u_r is the call recipient's utility. Any cost-based termination price mechanism can be optimal from a social point, as far as allocative efficiency is concerned, only if $u_r = 0$.[42] So, the CPP mechanism underlies the economic indifference of the party receiving the call regarding the termination price paid by the caller. This further reinforces the termination bottleneck.

[41] Art 75 EECC

[42] De Graba (2003); Berger (2005); Armstrong and Wright (2007).

In such a situation, even setting a perfectly cost-oriented termination price will never lead to an efficient allocation of resources. In theory, the direct and most appropriate solution to solve the problem of externalities would be to modify the structure of the retail price, introducing a *Receiving Party Pays* (RPP) system that efficiently allocates the costs of the information exchange.[43] In an RPP system, like the one adopted in the US, both customers, the caller and the receiver, pay their network operator to support the call's costs. In other words, the caller pays to originate the call and the receiver pays to receive it. The RPP model completely removes call externalities where both the caller and the receiver pay according to the benefits deriving from the call and, consequently, the respective retail prices determine the economic choices of the customers. This model solves the problems of termination bottlenecks since the network that terminates the call offers its services only to its customers and the demand for these services can react accordingly.

The adoption of an intermediate commercial model, the *Bill and Keep* arrangements, could have been an efficient solution. Bill and Keep is based on the CPP retail commercial model. However, it is associated with a system where all wholesale transactions between operators, entail zero termination prices. The adoption of a Bill and Keep system would have solved the economic problems associated with the current wholesale termination remuneration model.[44] At the same time, the end-user would not have undergone through a change in the retail commercial model.

References

AGCOM. (2019). Report on Online Platforms.

Aghion, P. et al. (2005). Competition and Innovation: An Inverted-U Relationship, Quarterly Journal.

Armstrong, M., Doyle, C., & Vickers, J. (1996). The Access Pricing Problem: A Synthesis. *Journal of Industrial Economics, 44*, 131–150.

Armstrong, M., & Sappington, D. (2006). Regulation, Competition and Liberalisation. *Journal of Economic Literature, 32*(2), 353–380.

Armstrong, M., & Wright, J. (2007). Mobile Call Termination. MPRA Paper N. 4858.

Arrow, K. J. (1962). Economic Welfare and the Allocation of Resources for Invention, in the Rate and Direction of Economic Activities.

[43] Littlechild (2006).

[44] De Graba 2003; Berger 2005; Armstrong and Wright (2007).

Averch, H., & Johson, L. (1962). Behavior of the Firm Under Regulatory Constraint. *Am Econ Rev, 52*(5), 1053–1069.

Bacache, M., Bourreau, M., & Gaudin, G. (2014). Dynamic Entry and Investment in New Infrastructures: Empirical Evidence from the Fixed Broadband Industry. *Review of Industrial Organization, 44*, 179–209.

Baker, J. B. (2007). Beyond Schumpeter vs. Arrow: How Antitrust Fosters Innovation. *Antitrust Law Journal, 74*, 575–602.

Berger, U. (2005). Bill & Keep vs. Cost-Based Access Pricing Revisited. *Economics Letters, 86*(1), 107–112.

Bouckaert, J., Dijk, T., & Van Verboven, F. (2010). Access Regulation, Competition and Broadband Penetration. *Telecommunications Policy, 34*, 661–671.

Bourreau, M., Cambini, C., & Hoernig, S. (2015). Geographical Access Markets and Investments in Next Generation Networks. *Information Economics and Policy, 31*, 13–21.

Bourreau, M., Doan, P., & Manant, M. (2010). A Critical Review of the "Ladder Investment" Approach. *Telecommunications Policy, 34*(11), 683–696.

Briglauer, W., Ecker, G., & Gugler, K. (2012). Regulation and Investment in Next Generation Access Networks: Recent Evidence from the European Member States. Working Papers, Research Institute for Regulatory Economics.

Cambini, C., & Jiang, Y. (2009). Broadband Investment and Regulation: A Literature Review. *Telecommunications Policy, 33*(10-11), 559–574.

Cave, M. (2006). Encouraging Infrastructure Competition Via the Ladder of Investment. *Telecommunications Policy, 30*(3–4), 223–237.

Cave, M. (2014). The Ladder of Investment in Europe, in Retrospect and Prospect. *Telecommunications Policy, 38*(8–9), 674–683.

Cave, M., & Vogelsang, I. (2003). How Access Pricing and Entry Interact. *Telecommunications Policy, 27*(10-11), 717–727.

Communications Committee. (2014). Working Document: Broadband access in the EU-COCOM 14-03.

Curzon Price, T., & Walker, M. (2016). Incentives to Innovate vs Short-term Price Effects in Antitrust Analysis. *Journal of European Competition Law & Practice, 7*, 475–482.

De Graba, P. (2003). Efficient Interconnection Rates for Interconnected Competing Networks. *Journal of Economics Management Strategy, 12*, 207–230.

European Commission. (2018). Digital Scoreboard.

European Commission. (2019). Study on Broadband Coverage in Europe 2018.

European Regulators Group. (2004). ERG Common Position on the Approach to Appropriate Remedies in the New Regulatory Framework. ERG (03) 30rev1.

Grajek, M., & Roller, L. H. (2012). Regulation and Investment in Network Industries: Evidence from European Telecoms. *Journal of Law and Economics, 55*(1), 189–216.

Hoernig, S. (2007). On-net and Off-net Pricing on Asymmetric Telecommunications Networks. *Information Economics and Policy, 19*, 171–188.

Kerber, W. (2017). Competition, Innovation, and Competition Law: Dissecting the Interplay MAGKS Joint Discussion Paper Series in Economics, 42-2017.

Laffont, J. J., & Tirole, J. (2000). *Competition in Telecommunications.* Cambridge: MIT press.

Littlechild, S. (1983). Report to the UK Secretary of State—Regulation of British Telecommunications' Profitability.

Littlechild, S. (2006). Mobile Termination Charges: Calling Party Pays Versus Receiving Party Pays. *Telecommunications Policy, 30*(5-6), 242–277.

Manenti, F., Distaso, W., & Lupi, P. (2006). Platform Competition and Broadband Uptake: Theory and Empirical Evidence from the European Union. *Information Economics and Policy, 18*, 87–106.

Odale, A., & Padilla, J. (2004). From State Monopoly to the "Investment ladder": Competition Policy and the NRF, LECG Europe, The Pros and Cons of Antitrust in Deregulated Markets, 51-77.

of Economics, 120, 2, 701-728.

Sappington, D., & Weisman, D. (2010). Price Cap Regulation: What Have We Learned from 25 Years of Experience in the Telecommunications Industry? *Journal of Regulatory Economics, 38*, 227–257.

Schumpeter, J. A. (1934). *The Theory of Economic Development.* Cambridge: Cambridge University Press.

Schumpeter, J. A. (1942). Capitalism, Socialism, and Democracy.

CHAPTER 6

(Ultra)Broadband Policies in Europe

Abstract The development of Next Generation and Very High Capacity Networks has become one of the main objectives of the telecom regulatory framework, as defined in the European Electronic Communications Code. Indeed, the huge capital-intensive investments, necessary to reach the most recent European and national policy targets, require a new regulatory balance between competition and innovative investments is not found. This in turn may imply an undesirable (and unintended) softening of competition dynamics and its tension toward dynamic efficiency. So far, responses to this difficult trade-off have emerged both from the market and from policymakers with new business models, market strategy (e.g. cooperative investments and network sharing) and regulatory approach (e.g., varying regulation intensity according to the risk of new investments). Finally, in most EU countries, a direct public intervention, aimed at financing the most innovative investments, has been put in place and needs to be reconciled with existing state aid rules and competitive dynamics.

Keywords Next generation networks • Very high capacity networks • Co-investment • Wholesale-only • Risk regulation • State aid

A. Manganelli, A. Nicita, *The Governance of Telecom Markets*,
Palgrave Studies in Institutions, Economics and Law,
https://doi.org/10.1007/978-3-030-58160-2_6

6.1 Next Generation Networks and the Evolution of Regulatory Approach

The 2002 New Regulatory Framework (NRF) focused on opening national telecom markets for the transition from a monopolistic to a competitive liberalised regime. The European Commission and National Regulatory Authorities (NRAs), through access regulation and the "ladder of investment" approach, promoted infrastructure-based competition among vertically integrated operators in order to have a sustainable and dynamically efficient competition at retail level. This was done by setting the menu of prices to the different access services for the incumbents' networks so as to encourage entrants to gradually build (some parts of) their own network. Technological innovation and, namely, the *Next Generation Networks* (NGN) in optic fibre and the consequent development of ultra-broadband digital services, began changing the economic, business and policy scenario of the telecom market.

Consequently, the public policy debate began to gradually focus on industrial policy issues, such as the transition from copper to fibre and the extensive development of future-proof technological assets, rather than solely on the regulation of market failures.[1]

Optic fibre networks had entirely replaced copper in the telecom network core backbones (long distance transmission) already twenty years ago. The main market and policy issue concerned and still concerns the access network (from local exchange to end-users' premises), where fibre is slowly and costly getting closer to the end-users. To simplify, there are three access network architectures (see Fig. 6.1):

(a) an entirely copper access network (ADSL/VDSL);
(b) Fibre To The Cabinet (FTTC), where the copper is limited to the segment between the street cabinet and end-user premises;
(c) Fibre To The Home (FTTH), an end-to-end fibre network.

Moreover, in quite a few European countries (with some exceptions as in the case of Italy and Greece), coaxial lines have been deployed by tv cable operators. There has been high heterogeneity in the deployment of Next

[1] Development of competitiveness also relies on the resolution of a specific market failure, i.e., the complex argument of positive externalities associated with broadband and ultra-broadband deployment.

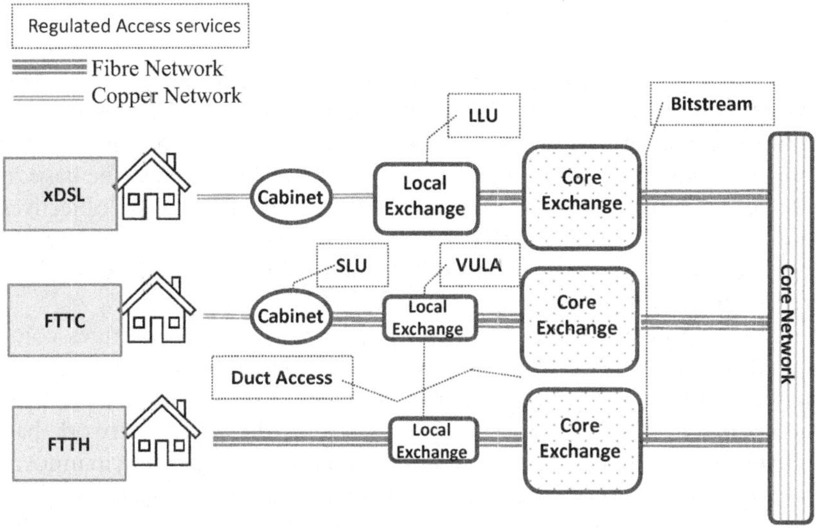

Fig. 6.1 Simplified access network architectures and regulated access services

Generation Access Networks (NGAN) in different countries, depending on the market situation (e.g., the existence of tv cable operators) and the regulatory approach.[2] Notwithstanding, co-existence of legacy copper and next generation access networks has been a common situation in many countries. In such cases, the two networks are strongly interdependent and consequently market strategies, economic incentives and public regulation are dealing with these complex interdependences.

The EU and the Member States progressively designed industrial policies for NGAN investments, by setting increasingly ambitious connectivity and broadband targets. At the European level, these industrial policies have been raised from the challenging yet achievable goals established within the 2010 Digital Agenda for Europe (DAE)[3] to the extremely ambitious objectives within the 2016 Gigabit Society (GS).[4] In a nutshell, the DAE asked Member States and undertakings to reach, by 2020, a universal availability of at least 30 Mbps for all EU households and 100 Mbps for 50% of EU households. The GS overall goal is to ensure, by

[2] Shortall and Cave (2015).

[3] European Commission (2010).

[4] European Commission (2016).

2025, a Gigabit connectivity (i.e., 1 Gbps = 1000 Mbps) for all main socio-economic drivers (e.g., education, government services, transport hubs, digitally intensive enterprises) and a universal access to at least 100 Mbps, extendable to 1 Gbps, for all European households, even in rural and remote areas. Those targets are not compelling, as communications are not binding. However, these policy objectives have been at the basis of recent legislative evolution, i.e., the EECC, both in terms of objectives and specific provisions. Moreover, targets also influenced the implementation of state aid rules, by means of updating the sector-specific guidelines reflecting the policy objectives (see Box 6.1).[5]

Such highly ambitious objectives implies regulatory challenges compared to the previous scenario. The main challenges are: (a) paramount new investments for infrastructures are needed, while in the previous access-based scenario, sunk costs for building the legacy network had already been sustained (in a monopolistic regime); (b) investing in innovative networks in an open market is much riskier, as the associated services are characterised by a high level of demand uncertainty due to (b_i) increased downstream competitive pressure (where there are more than one operator, so-called "black areas") and (b_{ii}) uncertainty about consumers' preferences and willingness to pay for new services; and (c) the ambitious GS NGAN coverage targets require material investments also in market areas where infrastructure competition and infrastructures duplication may not be feasible and/or desirable (so-called "grey areas") and even in areas where neither a provider is operating, nor are there plans to operate in the next few years (so-called "white areas").[6]

In order to pursue such highly ambitious objectives and address their challenges, new policy and regulatory approaches were developed, both amending the regulatory framework and using the existing regulatory tools in a different way. Overall, this new approach contributed to progressively reshape the balance between innovation and competition.

The main aspects of this adjustments mainly concerned:

- Indirect intervention, by means of regulatory tools and incentives (described in this section):

[5] Parcu and Rossi (2020).

[6] The *coloured area* approach has been developed in the context of the broadband guidelines on state aid (see Box 6.1).

  - Access price for the legacy network,
  - Specific regulation for next generation access networks,
  - Infrastructure sharing,
  - Access to physical passive civil infrastructures,
  - Symmetric regulation.
- Direct intervention (described in Sect. 6.3):
  - Public subsidies to the supply-side,
  - Demand-side policies.

In September 2010, the Recommendation on the Next Generation Access Network (NGAN)[7] was adopted to encourage timely and efficient investments and innovation, while safeguarding effective competition. In particular, the Recommendation confirmed that unbundling obligations should apply also to fibre network local access (VULA) in order to guarantee competition. However, a relaxation of the access regulation should have taken place in the case of alternative infrastructures or joint deployments of fibre to the home (FTTH) lines. Accordingly, access price had to be subject to cost-orientation adjustments, while a risk premium, together with pricing flexibility, would have ensured an adequate rate of return (e.g. price differentiation in case of long-term contracts and volume discounts).

In 2013, a Recommendation on non-discrimination and costing methodologies[8] was adopted in order to facilitate a smooth migration from copper to NGA and to provide incentives for pan-European network and services development. The Recommendation was designed to reach the following goals: (a) ensuring a non-discrimination environment by granting competitors an "equivalent" access (applying the same prices and processes to NGA wholesale products for competitors and for incumbent retail businesses, *Equivalence of Inputs*);[9] (b) keeping wholesale copper access prices at a stable level during the transition to NGA; (c) not

[7] European Commission (2010) Recommendation on regulated access to Next Generation Access Networks (NGA)—2010/572/EU.

[8] European Commission (2013) Recommendation on consistent non-discrimination obligations and costing methodologies to promote competition and enhance the broadband investment environment—C(2013) 5761 final.

[9] If EoI is disproportionate considering national circumstances at least an "Equivalence of Output" is required.

imposing price regulation (cost-oriented wholesale access pricing obligations) on incumbent operators' fibre-based NGA networks, in the presence of specific conditions providing alternative competition safeguards.

In 2014, the Cost-reduction Directive[10] was adopted to reduce costs for high-speed network deployment. The Directive focuses on four main areas: (a) obligations for owners of existing physical "passive" infrastructures in all sectors (e.g. ducts, poles or masts) to provide access under fair and reasonable terms in favour of access-seekers intending to roll out high-speed broadband networks; (b) coordination mechanisms for civil works in order to stimulate cross-sectoral synergies related to physical engineering works which may significantly reduce the need and cost for civil works; (c) streamlining of local administrative procedures for permissions and authorisation of works; (d) obligations for new buildings and major renovations to install infrastructures enabling and facilitating high-speed internet access (e.g. mini-ducts, access point) and access to in-building infrastructures.

In September 2016, within the so-called "Connectivity Package" for the Digital Single Market (DSM),[11] the EU Commission elaborated a proposal concerning an overall review of the Telecoms Regulatory Framework. Finally, in December 2018, the EU Electronic Communications Code (EECC) Directive was finally adopted, entailing an overall review of the Telecoms Regulatory Framework.

The 2018 updated regulatory framework represents the latest step in the evolution of the balance between the different regulatory objectives, namely competition enhancement and investment promotion. Even though the promotion of investments has always been one of the objectives of the telecom regulatory framework, its configuration varied and has become increasingly central.

With the 2002 NRF, promotion of investments was considered as a tool to foster sustainable and durable competition (infrastructure-based).[12] With the entry into force of the 2009 Better Regulation Directive,

[10] Directive 2014/61/EU.

[11] Besides, the EU Electronic Communications code, the other components of the package are: (a) New 2025 Gigabit broadband objectives: Commission Communication (COM(2016) 587 final); (b) The 5G Action plan: Commission Communication (COM(2016)588); (c) Wi-Fi for Europe: targeted voucher scheme: Regulation amending CEF.

[12] Art. 8(2)c FD: "The national regulatory authorities shall promote competition … by inter alia … encouraging efficient investment in infrastructure and promoting innovation".

investments were considered as key enablers of efficient and effective regulation.[13] Finally, within the 2018 EECC, investment promotion has become a new independent general objective of the European strategy. Accordingly, NRAs, as well as BEREC, the European Commission and the Member States were mandated to "promote connectivity and access to, and take-up of, very high capacity networks [VHCN], including fixed, mobile and wireless networks, by all citizens and businesses of the Union".[14] Consequently, competition promotion, including infrastructure-based competition, and the extensive deployment of VHC fixed and mobile networks have become two equally ranked objectives, which must be balanced by NRAs within their activity.

In particular, there are some specific EECC provisions that NRAs have to apply and shape this new balance between competition enhancement and investment promotion:

(a) imposition of obligations on undertakings to meet reasonable requests for access to civil engineering assets (such as duct access) that directly support competitive infrastructure deployment: this can be imposed as a self-standing remedy to be considered before assessing the need to impose any other potential remedies and regardless the fact the assets are part of the relevant market analysed in the procedure;[15]
(b) extension of the scope for access regulation to be potentially imposed on all operators, and not only on SMPs ("symmetric regulation") and also owners of wiring and cables who are not electronic communications network providers, when replication of network elements is "economically inefficient" or "physically impractical";[16]

[13]Art. 8(5)d FD: "The national regulatory authorities shall, in pursuit of the policy objectives … apply objective, transparent, non-discriminatory and proportionate regulatory principles by inter alia … promoting efficient investment and innovation in new and enhanced infrastructures, including by ensuring that any access obligation takes appropriate account of the risk incurred by the investing undertakings and by permitting various cooperative arrangements between investors and parties seeking access to diversify the risk of investment, whilst ensuring that competition in the market and the principle of non-discrimination are preserved".

[14]Art 3 (2) EECC.

[15]Art 72 and recital 187 EECC.

[16]Art 61(3,4) and recital 154 EECC.

(c) providing a better remuneration of investments ("reasonable rate of return") and adequate consideration of investment risk, especially with regards to VHCN, when regulators decide whether to impose access obligations and define access pricing;[17]
(d) permitting various cooperation arrangements between investors and access-seekers in order to share investment costs and diversify the investment risk, while ensuring that competition and non-discrimination safeguards are preserved;[18]
(e) promoting co-investments in VHCN by de-regulating (removing all access obligations) SMP operators who committed to an open offer for co-investments on FRAND terms (while also allowing access to non-co-investors);[19]
(f) granting a lighter regulatory treatment (i.e., less demanding access obligation) for wholesale-only network operators (meant as companies not active in the downstream retail market[20]) that enjoy a SMP position in the wholesale/upstream market, due to the smaller risk of anticompetitive discriminatory behaviours (compared to a vertically integrated SMP operator).[21]

Table 6.1 below provides with a useful systematisation of the EECC regulatory obligations, classified according to the different kind of operators to which these obligations could be imposed.

From the table, the different intensity of regulation obligations borne by different kind of operators can be perceived. The higher degree of deregulation, almost complete, is allowed to SMP operator engaging in VHCN co-investments under art 76 EECC.

Some clarifications for the "yes but" in the Table 6.1 are the following:

- Symmetric regulation for wholesale-only operators are possible, but not beyond the first concentration point, if operators are privately financed and offer an effective alternative access on a commercial basis to a very high capacity network, on fair, non-discriminatory and reasonable terms and conditions. There are no symmetric regulation exemptions for publicly financed wholesale-only operators;[22]

[17] Art 3(4), art 73 and 74 EECC.

[18] Art 3(4) EECC.

[19] Art 76 and 79 EECC.

[20] Wholesale-only operator can provide retail services to business users larger than small and medium-sized enterprises. Recital 208 EECC.

[21] Art 80 EECC.

[22] Recital 155 EECC.

Table 6.1 Type of operators and possible regulatory obligations

	SMP operator	*SMP VHCN Networks*	*Wholesale-only SMP operator*	*SMP VHCN co-investments*	*non-SMP operator*
Interconnection obligations (art. 60)	Yes	Yes	Yes	Yes	Yes
Symmetric regulation (art. 61.3)	Yes	Yes	Yes but (see below)	Yes	Yes
Transparency (art. 69)	Yes	Yes	No	No	No
Non-discrimination (art. 70)	Yes	Yes	Yes	No	No
Accounting separation (art. 71)	Yes	Yes	No	No	No
Civil engineering access (art. 72)	Yes	Yes	No	No	No
Access obligations (art. 73)	Yes	Yes but (see below)	Yes	No	No
Cost oriented Price (art. 74)	Yes	Yes but (see below)	No	No	No
Fair and reasonable prices (art 80.2)	No	No	Yes	No	No
Functional separation (art. 77)	Yes but (see below)	Yes but (see below)	No	No	No
Voluntary separation (art. 78)	Yes	Yes	No	Yes	No
Notification of migration (art 81)	Yes	No	Yes	No	No

- Access and cost-oriented price remedies for SMP operator investing in VHCNs are possible, but, as mentioned, the EECC requires NRAs to regulate them targeting a "reasonable rate of return" on operators' investments, under an adequate consideration of their investment risk;[23]
- Vertical functional separation of the incumbent operator (described in Sect. 2.3) is a very intrusive regulatory remedy, which has never been adopted in Europe as such. Although, the EECC has maintained

[23]Art 3(4), art 73 and 74 EECC.

such a remedy, given the new policy context focused on investments, an unprecedented vertical separation obligation for an SMP operator seems even more unlikely than before, especially in the current context where infrastructure competition has been strengthened. Differently, functional separation has been made by the EECC explicitly not possible for SMP engaging in VHCN co-investment (while binding commitments in a voluntary separation process, under art. 78, are still possible). All vertical separation remedies are obviously not for wholesale-only operators as they are already vertically disintegrated).

6.2 A Composite Market and Policy Framework

The definition of the balance between competition enhancement and promotion of investments has always involved the definition of access regulatory remedies (as detailed in Sect. 5.3). From the initial setting to the new EECC, the evolution of access regulation has been substantial, due to a few important reasons.

Firstly, an extended two-legged ladder including the NGAN is much more complex than a "traditional" one (See Fig. 5.1 for a comparison). Additional infrastructure-based access options have developed, such as sub-loop unbundling (SLU), where the access seeker place its equipment at the street cabinet, and the Virtual Unbundling Local Access (VULA) in the fibre networks (Fig. 6.2). The "vertical" policy objective of a new entrant progressively climbing the ladder, to develop infrastructure-based competition, is often associated with a "horizontal" policy objective of

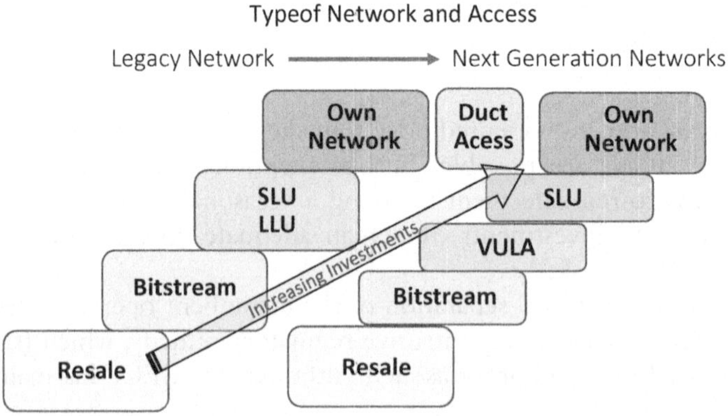

Fig. 6.2 The extended ladder of investments for legacy-NGA networks

incumbent transition from the legacy network to the next generation one. This create strong interdependences between operators and their incentives, which makes very difficult for regulators to set access charges in order to provide the right incentives to both boost investments and promote competition.

Therefore, another important objective for the definition of access charges could be to provide incentives for the transition to fibre. In this regard, the main policy question is whether access price to copper (the LLU price) can influence, i.e., accelerate, the transition to fibre.

The effect of a lower price of LLU on investments in fibre coverage is ambiguous, due to 3 conflicting economic effects:[24] (a) "replacement effect" for a new entrant, as profits earned by accessing incumbent's copper network represent for new entrants an opportunity cost to invest in fibre; (b) a "wholesale revenue" effect for incumbent which faces an opportunity cost of investing in fibre because of the loss of access revenues from copper; (c) a "retail migration" effect for both incumbent and new entrants. This last effect is the most complex: a lower copper access price would result in lower retail price, influencing retail prices of substitutes fibre-based services; in turn, this would ultimately affect fibre profitability and the relative incentive to invest. To sum up, a lower LLU access price in the legacy network would result in (a) stronger replacement effect and less investments from the entrant; (b) lower wholesale revenue effect and more investment from the incumbent; (c) a stronger migration effect and less investment for both the entrant and the incumbent. Therefore, an entrant would respond to a higher LLU price with more investment, whereas the effect for the incumbent is uncertain due to two conflicting effects. This last point has been clarified, verifying that retail migration effect is stronger than the wholesale revenue effect, considering competition with new entrants. Therefore, access regulation on legacy network affects incumbent investments on NGN, in a negative way, i.e., stricter access regulation, in terms of a lower LLU price as well as a larger scope of NGAN regulation, gives disincentives to incumbent's investment in NGN.[25]

A second crucial aspect of access regulation evolution relates to the emergence of a new crucial market and regulatory dimension, previously

[24] Bourreau et al. (2012).

[25] Briglauer et al. (2018).

not covered by the LoI approach. In addition to downstream competition (i.e. service-based by a vertically separated competitor versus facility-based competition by an integrated one), market strategies and public policy focused on upstream dynamics, i.e., upstream competition by wholesale-only operator versus upstream cooperation by companies engaging in infrastructure and risk sharing agreements.

A policy taxonomy of the market and policy framework is presented at Table 6.2 below, which considers the vertical structure of the incumbent's competitors (i.e., integration vs separation) and the type of competition they give rise to (i.e. service based vs facility-based). As described in the previous section, the LoI approach was intended to intensify downstream competition in the market, moving from service-based to facility-based, while the incumbent's competitors progressively changed their vertical structure from separation to integration (thus, climbing the rungs of the ladder and moving diagonally in the taxonomy, from quadrant I to quadrant II). Indeed, within the LoI approach, there was a correlation between the type (and intensity) of competition and the vertical structure of competitors. The incumbent faced service-based competition from vertically separated competitors, and facility-based competition from vertically integrated ones.

Table 6.2 A market and policy taxonomy

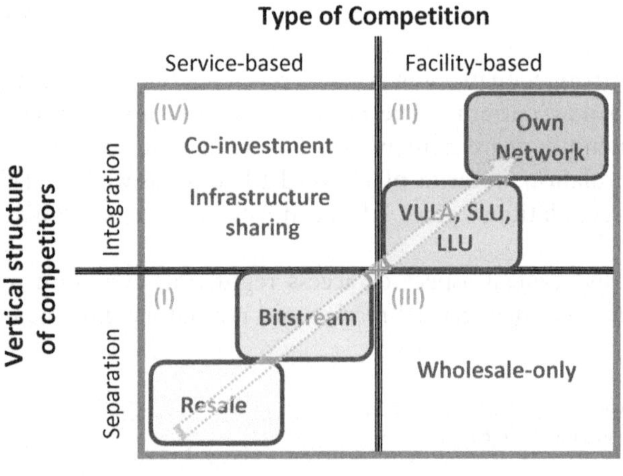

Of course, a continuum exists between separation and integration, which is represented here in a simplified dichotomy.[26] In any case, the need of new paramount network investments, and the consequent co-evolution of markets and rules, significantly increased the importance of market dynamics at upstream level, where competitors' choices are no longer related only to whether to *make* (by themselves) or to *buy* (from the incumbent) in order to compete downstream. This resulted in the extension of the possible combinations between different types of competition and the vertical structure of competitors (thus making quadrants III and IV relevant).

Market dynamics and ultra-broadband policy targets set the conditions for market entry of a few wholesale-only vertically separated operator (quadrant III). Consequently, an incumbent began experiencing a geographically differentiated upstream (facility-based) competition for the provision of network access to its downstream competitors (and to its downstream units). In turn, alternative downstream operators enjoyed an extended choice, as they could choose whether to make, or to buy from the incumbent (which is a competitor downstream), or to buy from an alternative upstream operator that is not a downstream competitor. This extension can make a big difference in terms of competition intensity, due to the creation of (a) upstream wholesale (more or less) competitive markets which could eventually imply a deregulation for the incumbents, and (b) competition safeguards for downstream competitors as wholesale-only access providers would have no incentives to discriminate among its clients. For the latter reason, the EECC included provisions granting a relaxation of regulatory obligations of SMP wholesale-only operators, proportionally tailored to the smaller risk and incentive that a wholesale-only operator has for a discriminatory treatment of its downstream business customers. Of course, the application of this rule would imply that the vertically integrated historical operator has been marginalised in the market, which is unlikely to happen at the national level. However, such an outcome is not unrealistic as far as local geographical markets are concerned. In facts, this outcome might result from a selective entry and investment strategy of a wholesale-only operator either driven by market profitability considerations (a demand-driven *cream-skimming*) or instead driven by public interest and public subsidies (a supply-driven approach; on this point see the following Sect. 6.3).

[26]See Cave (2006).

In case of investments limited to "white areas", i.e. to areas where a market failure occurs, the wholesale-only model represents a geographic complement of the Ladder of Investments approach; whereas if it has wider dimension, covering competitive areas where a LoI is in place, the wholesale-only model may show some disruptive features and may act in discontinuity with the existing LoI approach. Indeed, within the current regulatory path, a wholesale-only model has typically a local dimension. However, when a wholesale only operator aims at supplying a nationwide offer to access-seekers, it may affect their incentives to vertically climb the ladder of investments. As a consequence, a nationwide wholesale-only operator may imply the reduction or even the elimination of infrastructure-based competition developed by alternative operators. This is because a wholesale-only operator: (a) exercises (potential) competition on all (prospectively) vertically integrated subjects active in the upstream market; (b) alters the "make or buy" choice for the new entrant, not allowing regulators to incentivise competitors' infrastructure developments by modulating access prices to the incumbent's network. On the other side, a nationwide wholesale-only operator may nonetheless accelerate upstream investments even when the actual demand for ultrafast broadband services is fairly limited. Non-SMP wholesale-only operators, with different source of financing and investment strategies, are active in many EU countries: the largest coverage with a mix of public and private financing have been currently reached in Italy and France.

At the opposite side of the taxonomy (quadrant IV), forms of upstream cooperation between vertically integrated operators have become increasingly frequent as players seek to share costs and risks arising from infrastructure development. These forms of upstream cooperation involve downstream competition between companies that are already vertically integrated (typically the incumbent) or would become (more) vertically integrated following the cooperation. Due to investment facilitation and the almost complete deregulation of the VHCN SMP operators, as defined by the EECC, additional forms of cooperation between firms are likely to emerge. Differently from the wholesale-only model, co-investment and infrastructure sharing can be considered a cooperative evolution of the LoI driven by market demand and economics. In fact, many operators have engaged in sharing infrastructures and co-investment as a further step of a progressive development of competing infrastructures.

Here the main policy question is whether the new regulatory scheme of co-investment gives more incentives to roll-out of very high capacity networks compared to access regulation, which is the counterfactual of any assessment about co-investment and network sharing involving an SMP operator. Recent studies assess a market situation where a new operator can enter local areas, via standard access, paying a variable access price, and/or sharing incumbent's infrastructure costs.[27] Co-investment implies a higher network coverage, black areas expand, and stimulates competition in areas with service-based competition, leading to lower prices for consumers. Furthermore, white areas shrink, reducing the extension of areas in needs of public subsidies (see next Sect. 6.3). Yet end-users in grey areas may experience higher prices. Overall effect on welfare is positive.

A particular aspect of open co-investments concerns uncertainty about demand and investment profitability and late entrants, who could play opportunistically under an obligation to keep the offer to co-invest open to any provider at any moment during the lifetime of the network.[28] If entrants can wait to co-invest until demand is realised, the incumbents' investment incentives are reduced and total coverage can be lower than compared to a counterfactual with earlier co-investment. This drawback can be remedied by a *risk premium* paid ex-post by entrants, as that one foreseen by the code,[29] although this would imply a lower entry.[30] As for the interplay between infrastructure sharing and other regulatory provisions, adding regulatory provisions that are substitutes for co-investment, such as access obligations, may promote opportunistic behaviours from entrants, reducing co-investment. By contrast, regulatory provisions that complement co-investment, such as access to ducts and poles, facilitate co-investment and improve coverage.[31]

The most problematic parts of both these new regulatory paradigm extensions are the interdependences between the "old" LoI approach to downstream competition (service vs facility-based: quadrants I and II) and the "new" upstream policy dimensions (competition vs cooperation: quadrants III and IV). Indeed, access regulation of incumbent networks (VULA-LLU-SLU) have countervailing effects on downstream and upstream competition. As for the wholesale-only model—in its more extensive version, i.e., covering competitive areas and interplaying with (more or less) vertically integrated competitors—its interdependences with

[27] Bourreau et al. (2018).
[28] Art 76 (1) a EECC.
[29] Art 76 (1) d EECC.
[30] Bourreau, Cambini, Hoernig, Vogelsang (2020).
[31] Bourreau, Hoernig, Maxwell (2020).

the traditional LoI regulatory approach create a regulatory trade-off between downstream and upstream competition. In a nutshell, at a given level of competitor infrastructure deployment, low access prices will intensify downstream competition. At the same time, this will discourage upstream competition by wholesale-only operators as incumbents' downstream competitors would prefer to buy from the incumbent, insofar as the (perceived) risk of discrimination is compensated by a lower price. Therefore, in a sense, the wholesale-only is a model with a pro-competitive impact, however in order to develop extensively (also in competitive areas) it must (temporarily) soften existing downstream competition in those areas. As for the co-investment model, if downstream competition has been a driver for investments, upstream cooperation (in the form of infrastructure sharing and co-investments) could short-circuit and jeopardise competition at the downstream level. That is why promoting cooperation and upstream de-regulation need adequate competition safeguards as those included in the EECC provision, complemented by supervision from competition authorities (see Sect. 3.3).

An additional variable which may further complicate this composite framework concerns the vertical structure of SMP operators. The above taxonomy at Table 6.2 refers to that kind of competitors and competition which a vertically integrated dominant incumbent would face. However, in quadrant IV, the relationship between cooperation and competition might at a certain point develop into a static form (and not dynamic as for the LoI), imploding into a more complex and strategic scenario, also involving a change in the incumbent vertical structure. Upstream cooperation necessary to intensify innovative investments could be so intense to be efficiently internalised into a single firm by means of mergers and acquisitions. In this case, vertical competition safeguards would become even more relevant (both as regulation and/or merger remedies). Within this conceptual approach, a voluntary functional separation remedy could acquire a new and strategic meaning for incumbents. In the different scenario of a wholesale-only operator (quadrant III) becoming an upstream dominant company, although there would be a lower risk of discriminatory practices, there would be still the need of independent ex-ante pro-competitive rules, i.e., price regulation, in order to avoid exploitative abuses.[32] Moreover, the latter scenario would

[32] An exploitative abuse of dominant position entails an exploitation of market power directly harming customers by also "directly or indirectly imposing unfair purchase or selling prices", art 102 (a) TFEU. Whereas within an exclusionary abuse, a dominant firm aims to strengthen or maintain its market power by putting rivals at a disadvantage, and thus may indirectly harm consumers.

likely have a long-term impact also on the vertical structure of alternative operators, refraining alternative infrastructural investments, thus decreasing retail service differentiation, and pushing toward a more intense downstream price-competition. An hybrid model in-between quadrant III and IV is also a possible scenario. In this regard, a model defined "Open Integrated Infrastructure", including both a vertically integrated incumbent and a wholesale-only opearator, has been recently proposed in Italy. The announced hybrid model seems to adopt features belonging both to the 'co-investment' one (i.e., including the possibility for third-parties to either co-invest or access the end-to-end infrastructure) and 'wholesale-only' one (i.e., a vertical separation and a governance design that effectively neutralise any vertical ties between upstream and downstream markets).

6.3 State Direct Intervention: Supply-Side Versus Demand-Side Policies

The EU regulatory framework is gradually including regulatory provisions aimed at matching industrial policy targets in terms of coverage of new generation access networks (NGAN) and adoption of ultrafast BB services, as defined by the DEA and GS.

Over ten years, from 2008 to 2018, telecom operators invested in Europe more than € 420 billion in telecommunication networks.[33] As described in the previous Sect. 6.2, despite the very composite framework and the different objectives to pursue, regulation have provided incentives to investment in new networks. Nevertheless, private investments are unlikely to be sufficient to match the EU policy targets of ultrafast BB coverage, considering the combinations of high deployment costs and demand uncertainty for such high-bandwidth access networks, especially in non-urban areas.

For this reason, in 2010, the DAE encouraged the Member States to develop national plans to stimulate broadband rollout through public funding from either national or EU resources. National broadband plans and strategies have been developed by almost all EU countries and public funds were allocated to finance deployment of NGAN networks. Those financing initiatives mainly focused on connecting rural areas. These measures, however, need to comply with EU state aid rules to be legitimately granted (see Box 6.1 below).

[33] ETNO (2019).

We can look at the evolution of NGAN coverage in the EU (Fig. 6.3 below), distinguishing between connections of at least 30 Mbps (fast BB) and at least 100 Mbps (ultrafast BB).

The growing EU coverage and take-up trends are evident. There is however a large heterogeneity across EU countries, in terms of (a) overall coverage, and (b) break down into different technologies and speed categories. The corresponding proportion of investment is very variable also within the overall NGAN category, comprising Fast BB and Ultrafast BB (Fig. 6.4 below). As a mere example, in 2019 France Ultrafast BB networks covered around 50% of households, slightly under the EU average, yet the total coverage of NGAN is 60%, much less than the EU average. On the other side, in Italy the overall NGAN coverage is about 90% of households, more than EU average, but the Ultrafast BB portion is much lower, standing at 25%.

Given the current coverage (at end of 2019), it has been estimated that further investment requirements to meet the DAE and GS targets amount to €384 billion, and that expected private investments would only cover up to one third of that amount.[34] This result obviously stresses that policy targets for VHC networks are far beyond what market forces can deliver

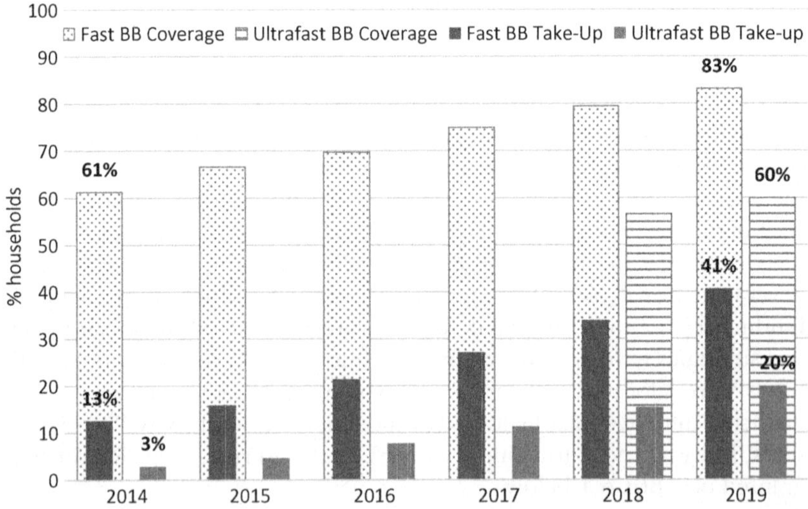

Fig. 6.3 Evolution of NGA coverage and take-up, EU average (Data in Figs. 6.3 and 6.4 comes from European Commission (2019))

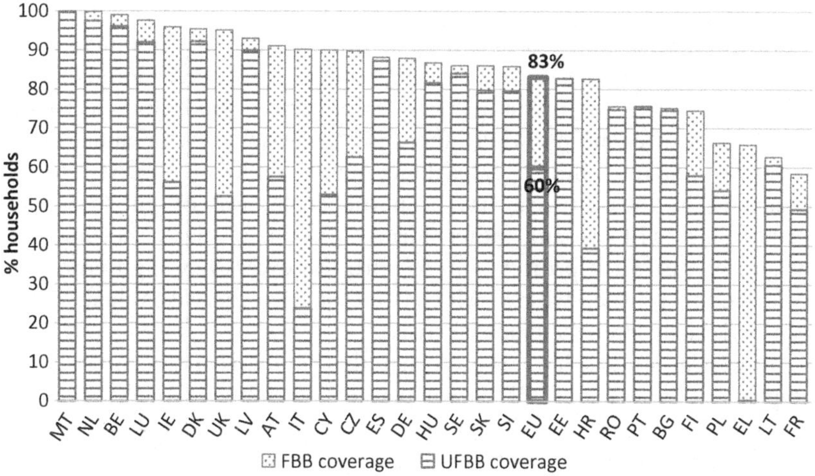

Fig. 6.4 NGA (fast BB + ultrafast BB) coverage in 2019, per MS (Data from European Commission 2019)

and calls for a substantial degree of public support to fulfil the investment gap, should Member States decide to realistically pursue those targets.

Of course, this is not a trivial policy decision for Member States, because it implies:

(a) re-allocation of public funds from other socially desirable objectives; and such re-allocation should prove efficient and socially beneficial;
(b) assessment of the demand for ultrafast BB services, aimed to ascertain whether (b_i) a significant portion of positive externalities are not reflected in the current demand level, or (b_{ii}) there is a willingness to pay unmet by current supply;[35]
(c) analysis of the interdependencies between public and private investments, as the latter can be unlocked by public funds but also crowded out, depending on the financing methods and the market characteristics of the subsidised geographic area (for example we have seen that co-investment could reduce the extension of "white areas") (see Box 6.1).

[34]Gruber (2019).
[35]Parcu and Rossi (2020).

EU funds could somehow soften the (a) dilemma. At the EU level, the most significant source of financing is the Regional Development Fund (ERDF), which is part of the European Strategic Investment Fund (ESIF). These funds support investments in "strategically important" sectors of the European economy, where there is otherwise a risk of underinvestment and market failure. Besides direct grants, a relevant approach to EU financing in the last year has been the facilitation of public-private-partnerships (PPPs) and other forms of public/private co-financing arrangements. Such projects aim to curb the excessively high enterprise risk faced by private investors which could hinder otherwise valuable projects, especially after the financial crisis.[36]

Empirical studies have shown that the emphasis in recent years on implementing forms of PPPs is well placed. The adoption of PPPs has had a positive effect on broadband penetration, especially in areas where a satisfactory level of competitive dynamics has not yet been achieved.[37] Furthermore, the European Investment Bank (EIB) is playing a key role by supporting the development of broadband infrastructure in Europe through so-called "soft loans" and other financial guarantees, as its main policy objectives.

Having said this, most analysists underline the discrepancies between the ambitious targets set by the EU and the small share of EU public funds dedicated to (ultra) broadband infrastructure: only € 3–4 billion annually of EU public expenditure,[38] against €250 billion of public funding by 2025 estimated as necessary to meet the GS targets.

As for (b), much less attention has been placed on the demand dynamics and on the role of public intervention in sustaining policies on the demand-side which may instead play a crucial role. As a matter of fact, a dynamic interplay between supply and demand exists, although it is unclear whether the demand for innovative services pushes the deployment of BB and ultra-BB networks, or, on the contrary, demand for innovative services is enabled by an adequate infrastructure (i.e., this is a classical *chicken-egg* situation).

This problem would be a no-brainer if there was an adequate demand in the market. However, as it can be seen at Fig. 6.3 above, the take-up and thus demand for ultrafast BB services, especially in some national

[36]Financial tools adopted are, for example, guarantees for losses, subordinated debt and other forms of "quasi equity" and equity finance arrangements, under which public authorities assume a much larger share of the financial risk than their private co-investors.

[37]See for example Belloc et al. (2012).

[38]Feasey et al. (2018).

markets, has proved to be relatively low compared to the coverage of the existing NGAN infrastructures. Economic analyses have shown that willingness to pay for innovative services is relatively low, in particular for very-high speed broadband services. This is in line with data that indicates the low adoption of ultrafast broadband services in most EU countries.[39] For this reason, demand-side policies have become more important and sensitive to the heterogeneity of new networks and services diffusion.

Some interesting suggestions for policy makers in this regard have been put forward, such as:

(a) setting adoption as well as coverage targets for ultrafast BB at both European and national levels;
(b) exploring the use of collective purchasing programmes, organised and administered by local authorities or non-profit bodies, to promote collective adoption of ultrafast BB in areas where such infrastructures have already been built (as well as to promote deployment in areas where they have not);
(c) revisiting the state aid guidelines on broadband networks in order to ensure that recipients of public funds have the appropriate incentives to meet ultrafast BB adoption targets, as well as roll-out targets;
(d) anticipating consumer future needs to design ultrafast BB *social tariffs* to support households unable to fund the ongoing costs of a UFB connection.[40]

Moreover, looking at past experiences regarding demand-side determinants of BB penetration, despite a different level of market maturity of the technology, some useful insights for the policy debate can be learnt. Over the past decades, key demand-side factors having fostered a capillary spread of broadband have been: (a) public demand for digital services; (b) incentives for the residential and business demand; (c) demand aggregation policies; (d) direct subsidies to consumers for terminals or service subscriptions.[41] As for the choice on the timing of supply-side and demand-side policies, the most effective actions have been those that provided a clear sequence of intervention: first, supply-side measures aimed at increasing investments and expanding supply capacity so as to achieve a minimum level of infrastructure; only afterwards, stimulation of demand.

[39] Briglauer and Cambini (2016).
[40] Bourreau et al. (2017).
[41] Belloc et al. (2012).

A comprehensive empirical analysis estimated the effect of each demand-side policy on broadband penetration, and found that supply-side policies (i.e., long-term loans to operators, national funding programmes, tax incentives to operators, PPPs, territorial mapping, administrative simplification) are important, but only for low and intermediate levels of broadband spreading. On the contrary, the impact of policies on the demand side is positively related to the degree of broadband diffusion (Fig. 6.5).

Albeit the backward-looking nature of this analysis, the bottom line is that supply-side and demand-side policies for the spread of broadband are complementary. An effective public intervention should therefore allocate resources on both sides, while ensuring a comprehensive overall long-term coordination. This is somehow confirmed by the observation of existing discrepancies between coverage and take-up of all broadband technologies and speed categories. In this context, vigorous demand-side policies should be implemented as soon as possible in order to boost access to existing network and facilitate the migration of users to new generation networks.

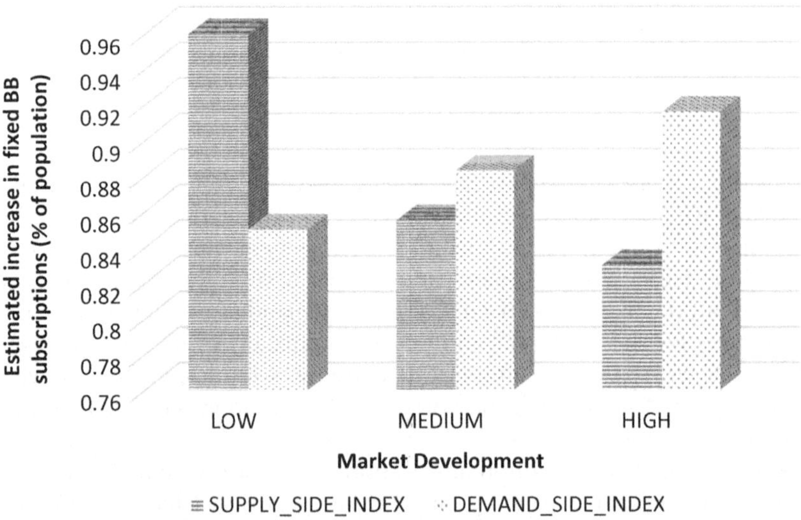

Fig. 6.5 Effectiveness and timing of demand-side vs supply-side policies (Data from Belloc et al. (2012))

Box 6.1 State aid rules for broadband: supply and demand-side aspects

The Treaty on the Functioning of the EU (TFEU) establishes a general prohibition on *state aid*.[42] In particular, state aids rules are based on the EU founding principle that state intervention should not distort market dynamics by favouring one competitor over other(s), crowding out existing or potential private investments. To qualify as state aid: (a) an intervention directly by the state or indirectly through state resources; (b) which has granted an advantage to the addressee on a selective basis; (c) must cause a (potential) distortion of competition; (e) possibly affecting trade between Member States. Nevertheless, legal framework allows some policy objectives to be pursued by means of the public financing of market players that qualify as state aids, provided certain conditions are met.

Consequently, in consideration of the growing pressure for public intervention in the telecom sector, the EC adopted specific *Broadband State Aid Guidelines* in 2009 and updated them in 2013.[43] The main principles were: (a) clear-cut distinction between basic and NGA networks; (b) technology neutrality; (c) effective access obligation to be imposed on networks receiving public subsidies regardless of any SMP position; (d) non-crowding-out (i.e., public subsidies must not be used in areas where private companies would find it profitable to autonomously invest, the so-called *black areas*);[44] (e) balancing state aid rule objectives with the broadband connectivity policy (i.e., regardless of the *colour* of the area, the public intervention may be allowed if it implies a "step change" in the supply-side of the market).[45]

(*continued*)

[42]Article 107 (1) of the TFEU.

[43]European Commission (2013).

[44]The *coloured area* approach comprises (a) white areas, where neither a provider is operating, nor are there plans to operate in the next three years. In white areas aids for both basic broadband and NGA networks are likely to be compatible with the internal market, as the aid promotes territorial cohesion and economic developments; (b) grey areas where only one provider is operating, and no other provider is likely to operate in the next three years. In grey areas aid must be subject to a full and detailed compatibility assessment, (c) black areas where there are or will be in the next three years, two or more operators providing basic broadband (coverage and infrastructure competition). In black areas aid is generally not allowed.

[45]A step change will result if public intervention implies: (a) significant new investment; and (b) enhanced capabilities, in terms of broadband service availability, capacity, speed and competition. In black areas, public investments for deployment of a new network will be considered

Box 6.1 (continued)

From 2009 (when the first guidelines were issued) to 2019, the European Commission took 125 decisions on state aid involving broadband and mobile connections, out of which only in one case the Commission found that the aid was not compatible with the common market and thus not put into effect.[46] A thorough revision of state aid rules is expected to take place in 2020, in order to consider the new EECC provisions, the EU funding programmes and 2025 policy objectives.

It is worth highlighting that some interesting amending proposals stress the importance of a goal-oriented application of state aid policy, also aimed to represent a (further) element of coordination between supply-side and demand-side policies. At the beginning, state aid had been deployed to extend broadband networks when the majority of households in urban areas (up to 80%) did not use the broadband services already available to them. A more balanced application of state aid policy should thus include elements regarding demand-side (such as adoption and coverage), in order to better pursue social inclusion objectives. For example, the extension of network coverage through state aid should be allowed only when adoption rates in the area at issue have reached certain thresholds. On top of this, these measures need to be coordinated with another demand-side policy, namely the Universal Service. In this case, the interaction of demand and supply-side mainly involves the affordability aspect. Current Broadband State Aid Guidelines stress that, according to the social inclusion objective, publicly subsidised broadband networks must be offered at the same prices as that charged in non-subsidised regions. However, within the Universal Service regime, this objective may be better pursued by directly subsiding budget-constrained households that otherwise would not be able to afford the fee for enjoying the infrastructure-based service supported by the state aid.[47]

a "step change" if all the following take place: (a) existing NGA networks, and those planned for the next 3 years, do not reach the end-user premises with fibre networks; (b) the market is not providing ultra-fast services above 100 Mbps in the near future, by the investment plans of commercial operators; (c) demand for such qualitative improvements is expected.

[46] Under art. 9(5) Regulation (EC) 2015/1589 laying down detailed rules for the application of Article 108 of the EC Treaty.

[47] Feasey et al. (2018).

References

Belloc, F., Nicita, A., & Rossi, M. A. (2012). Whither Policy Design for Broadband Penetration? Evidence from 30 OECD Countries. *Telecommunications Policy, 36*(5), 382–398.

Bourreau, M., Cambini, C., & Dogan, P. (2012). Access Pricing, Competition, and Incentives to Migrate from "Old" to "New" Technology. *International Journal of Industrial Organization, 30*(6), 713–723.

Bourreau, M., Feasey, R., & Hoernig, S. (2017). *Demand-Side Policies to Accelerate the Transition to Ultrafast Broadband—Project Report.* Cerre.

Bourreau, M., Cambini, C., & Hoernig, S. (2018). Cooperative Investment, Access, and Uncertainty. *International Journal of Industrial Organization, 56*(C), 78–106.

Bourreau, M., Cambini, C., Hoernig, S., & Vogelsang, I. (2020a). *Co-Investment, Uncertainty, and Opportunism: Ex-Ante and Ex-Post Remedies.* CESifo Working Paper Series 8078, CESifo Group Munich.

Bourreau, M., Hoernig, S., & Maxwell, W. (2020b). *Implementing Co-Investment and Network Sharing.* CERRE Report.

Briglauer, W., & Cambini, C. (2016). Determinants of Fast and Ultra-Fast Broadband Adoption, the Future of Broadband Policy: Public Target and Private Investments. In *The Future of Broadband Policy: Public Targets and Private Investment.* FSR-EUI.

Briglauer, W., Cambini, C., & Grajek, M. (2018). Speeding up the Internet: Regulation and Investment in the European Fiber Optic Infrastructure. *International Journal of Industrial Organization, 61*(C), 613–652.

Cave, M. (2006). Six Degrees of Separation: Operational Separation as a Remedy. *European Telecommunications Regulation, Communications & Strategies, 64,* 89.

ETNO. (2019). *The State of Digital Communication.* ETNO.

European Commission. (2010). Communication on A Digital Agenda for Europe—COM (2010) 0245.

European Commission. (2013). Communication: EU Guidelines for the Application of State Aid Rules in Relation to the Rapid Deployment of Broadband Networks (2013/C 25/01).

European Commission. (2016). Communication on Connectivity for a Competitive Digital Single Market—Towards a European Gigabit Society—COM (2016) 587 Final.

European Commission. (2019). Digital Economy and Society Index (DESI)—Connectivity Report.

Feasey, R., Bourreau, M., & Nicolle, A. (2018). *State Aid for Broadband Infrastructure in Europe.* CERRE Report.

Gruber, H. (2019). *Very High Capacity and 5G Networks: From the EU Code to the EU Market.* Presentation at DEEP-IN Workshop.

Parcu, P. L., & Rossi, M. A. (2020). State Aid Policy in the Broadband Sector: Public Announcements, Investments and Crowding Out. In P. L. Parcu, G. Monti, & M. Botta (Eds.), *EU State Aid Law* (pp. 99–120). Edward Elgar.
Shortall, T., & Cave, M. (2015). Is Symmetric Access Regulation a Policy Choice? Evidence from the Deployment of NGA in Europe. *Communications and Strategies, 98*, 17.

CHAPTER 7

The Evolution of Mobile Communications and Spectrum Policy

Abstract Over the last five years, there has been an enormously fast expansion of broadband services consumed in mobility. Extensive Machine-to-Machine (M2M) communications, also in the context of the so-called Internet of Things (IoT), have begun to make up an important share of all mobile communications. Furthermore, 5G, the latest mobile communications technology standard to enter the market, has emerged as a new way of providing mobile ultra-high capacity and very low latency connections, supporting a wide range of innovative applications and services. These phenomena are contributing heavily to the widespread use of data traffic, causing a paradigm shift in the existing economic patterns and rationales of mobile markets. All this has raised the need for updating some regulatory measures and has necessitated a deeper rethinking of spectrum policy strategies. The European Electronic Communications Code defines the current EU spectrum policy and is steering 5G spectrum auctions toward a more systemic approach.

Keywords M2M communications • Mobile broadband • 5G • Spectrum policy • Spectrum auctions

A. Manganelli, A. Nicita, *The Governance of Telecom Markets*, Palgrave Studies in Institutions, Economics and Law, https://doi.org/10.1007/978-3-030-58160-2_7

7.1 Mobile Broadband, M2M, IoT and 5G

Over the years, mobile communications have become widespread and, in 2002, overtook fixed-line services in terms of global subscriptions.[1] Mobile communications have had a very positive and lasting impact on economic growth and have often been the driver for public and private investments in broadband technologies. Mobile broadband has undergone a significant expansion in terms of subscriptions and traffic and evolved in a number of different ways.[2]

Mobile communications have become the main communication tools in today's economy and society, and this is not only for people. Our daily life has become populated with objects exchanging data and information by means of electronic communications technologies with limited human interaction, so-called Machine-to-Machine communication (M2M). Often these objects are connected through and with the Internet network and constitute the so-called Internet of Things (IoT).[3]

In the EU, and globally, we have seen a strong growth in object connections, driven by both the more traditional areas using the mobile cellular network, as well as the applications that use other communication technologies. In 2018, over 1 billion objects connected via the cellular network[4] were registered, a marked increase compared to the past. The total market value associated with M2M solutions came to over $250 billion in 2018, with $25 billion originating from connectivity.[5]

The connectivity model underlying the IoT has quite different characteristics compared to those related to traditional data applications. Firstly, IoT services need continually available connections *(always on)* and a ubiquitous connection (*everywhere*). Moreover, the global nature of the services, mainly based on the use of the mobile network, calls for the

[1] According to ITU statistics, in 2002, both world fixed and mobile lines amounted to around 1 billion, while in 2019, fixed-line subscriptions were approximately 930 million and mobile subscriptions were up to 8.3 billion.

[2] The mobile sector totalled over 6 billion broadband subscriptions in 2019, where fixed broadband connections were around 1.1 billion. See Ericsson (2019).

[3] There is a wide and diverse range of M2M and IoT services, including the smart home (domotics or home automation system, which can control a number of home appliances and systems; connected and autonomous cars; smart metering and the smart grid (electricity and gas sectors); telemedicine and digitised healthcare systems.

[4] Ericsson (2019).

[5] GSMA (2019).

development of connectivity offers that are not confined to national borders. These factors may create misalignments within the regulatory frameworks, requiring attention from policy makers to minimise risks of regulatory failures. A specific example concerns the so-called permanent roaming, namely the permanent presence in a country of devices with foreign mobile numbers (e.g., gas metering) where connectivity is established by a domestic operator but using a foreign EU SIM card. These services benefit from maximum wholesale prices under the EU regulation on international roaming, which has, however, been designed for the use of consumers who travel abroad and, therefore, for a temporary use. This situation calls for the identification of alternative solutions to reduce the related opportunistic behaviour, and create a level playing field for all national players.

The complexity of market changes related to the IoT scenario is augmented by two elements: (a) the formation of supranational alliances between national mobile operators in order to meet the global connectivity required, often resulting in different non-interoperable proprietary solutions; and (b) the evolution of business models and the growing number and type of relevant actors in the value chain. As for the latter, the IoT is pushing mobile communications from a traditional Business-to-Consumer (B2C) model towards Business-to-Business (B2B) or Business-to-Business-to-Consumer (B2B2C) models. In these new models, telecom operators are increasingly losing their direct relationship with the end-user, intermediated by other players and their businesses—the so-called *vertical industries*. These are digital platforms, automotive players, energy companies, healthcare companies, big industrial manufacturers and so on. For all the above reasons, a number of initiatives have been promoted at the EU and international level, involving policymakers, electronic communications operators and the vertical industries, in order to foster a sustainable and coordinated development of IoT services.[6]

These market and technological trends are heading towards what is called the "fifth generation" (or 5G) of telecommunications systems. 5G network architecture will result in the mobile network increasingly complementing the fixed-line network, finally achieving the long-debated fixed-mobile convergence, to deliver the required performance and capacity. 5G

[6]In 2015, the European Commission launched the Alliance for Internet of Things Innovation (AIOTI), to establish a competitive European IoT market and create new business models.

is supposed to be a building block in the forthcoming digital market society by supporting a wide range of innovative applications and services which will be pushing forward the industrial transformation (e.g. assisted driving, home automation, eHealth, energy management, possibly safety applications, etc.) and, in turn, provide further incentives for M2M and IoT

The 5G ecosystem involves the introduction of a meta-standard, connecting people and things, using the entire available spectrum, at various bandwidth levels, via interoperable modality, high capacity and very low latency.[7] Furthermore, the 5G standard will enable the full maturity of the *network slicing* approach, a new way of managing, operating and *orchestrating* multiple virtual dedicated networks in a common physical infrastructure.[8]

The EU 2025 GB Society Strategy has itself a broad vision of availability and take-up of very high capacity networks, wired and/or wireless. Along with Gigabit connectivity targets previously described, the new EU 2025 objectives include an uninterrupted 5G coverage for all urban areas and major terrestrial transport routes. The EU 5G Action Plan, launched in September 2016, proposes a framework for Member States and industry, for cooperation in and development of 5G wireless technologies through a public and private investment mix.[9] Many forms of public-private partnerships already exist in Europe,[10] involving large companies and universities. 5G deployment is a crucial policy and regulatory challenge for the coming years.

7.2 Law and Economics of Spectrum Management: The Need of a Systemic Approach

All mobile services described rely on the availability of radio spectrum, which is the basic resource asset for all wireless communications, including wi-fi and broadcasting. Consequently, the evolution of digital mobile services depends on how radio spectrum resources are accessed and managed.

[7] 5G should offer data connections well above 10 Gigabits per second, latency below 5 milliseconds and the capability to exploit any available wireless resources (from Wi-Fi to 4G) and serve simultaneously up to 1 million connected devices per km^2.

[8] GSMA, An Introduction to Network Slicing, 2017

[9] European Commission (2016).

[10] In 2013, the Commission launched the 5G Infrastructure Public Private Partnership (5G PPP), a joint initiative between the European Commission and European ICT Industry (ICT manufacturers, telecommunications operators, service providers, SMEs and research institutions).

Radio spectrum has been traditionally depicted as a scarce resource relative to some well-defined possible exclusive uses. It is a resource that belongs to the public dimension of State-owned assets, which can be assigned to private operators for business purposes, through a system of authorisations or licenses. Assignment of public resources to private actors must always have social objectives, which could be (a) indirect, i.e., receiving a fair remuneration for the private use of public assets and use those funds for social relevant activities; and (b) direct, by pursuing a socially desirable outcome in that specific sector, e.g., mobile service coverage and connectivity targets, which can have extensive positive externalities in the overall economy. The design of efficient market mechanisms could help in assigning this scarce resource, and its portions, in the most efficient and socially desirable way. [11]

This regulatory task is becoming increasingly difficult due to growing amount of spectrum needed for current mobile services. The progressive expansion of broadband services consumed in mobility by people, that require *everything and everywhere,* the progressive expansion of *always-on* M2M communications and IoT, the growing "spectrum hungry" applications and software, demanding ultra-wideband connection at mobile level, have implied an explosion of mobile data traffic. This is set to increase exponentially in the coming years,[12] and some analysts are afraid this might lead to a *spectrum crunch.* However, the risk magnitude of a "spectrum crunch" does not only depend on quantitative figures of spectrum avail able, but also on the quality of the available spectrum. In fact, there is a need for spectrum portions wide enough to support high data rate services with a predictable Quality of Service.

Another approach would suggest that spectrum resources are not scarce, per se, but finite, as the potentiality of non-exclusive uses, for instance through alternative sharing mechanisms, may in principle satisfy the entire market demand in an efficient way. Under this view, it would be the exclusive use of spectrum portions that exacerbates its scarcity features. This vision has clear merits; however, a difficult trade-off here lies with the fact many investments applications call for exclusive use of spectrum resources.

[11] See Pogorel (2018); Cave and Nicholls (2017).

[12] Ericsson estimates a +320% growth by 2025 up to 160 Exabyte/month, with traffic due to smartphones accounting for over 92% of total mobile traffic. See Ericsson (2019).

In any case, there is no doubt that an efficient release of spectrum, for example to enhance the mobile broadband and 5G communications, will not only depend on new spectral resources, but also on how the existing spectrum bands are managed along the full set of spectrum bands. Along these lines, in the last years the World Radiocommunication Conference in Geneva has repeatedly made it clear that an inefficient spectrum management is intolerable.[13] Indeed, the institutional design of spectrum policies has often been unable to ensure that spectrum supply adequately meets demand, and, today, this constitutes an incredible challenges for policymakers. As mentioned, spectrum governance must be efficiency-oriented also in order to pursue specific social objectives, e.g., aiming to achieve access to certain services for the greatest number of citizens-users.[14]

In order to promote an efficient use of spectrum, policymaking must consider and understand: (a) what the best uses are, i.e., the uses the maximise the economic and social value of the spectrum; (b) who will be willing to invest more in those resources, in order to implement those uses. In other words, what type of spectrum management design, in terms of *assignment* and *allocation*, would better perform. Allocation refers to the definition (initial allocation) or modification (so-called *refarming*) by the public authority of permitted uses or technology for which a frequency range is identified. Assignment refers to the identification of a specific holder for spectrum right of use. Assignment mechanisms, i.e., criteria and procedures under which frequencies are given, can be very different, using either (a) *market mechanisms*, e.g., auction, beauty contest and so on, or (b) *command & control* type-of mechanisms, where the public authority in charge discretionally identify one or more assignees.

Regardless the assignment mechanism, one of the main problems faced by policy makers or regulators is how to define those right of use, which in economic terms, are "property rights". As for all property rights, the main feature concerns whether the holder can "exclude" others from the use (thus having an exclusive right) or not. Another fundamental feature is related to the possibility of trading or leasing those rights, that is whether a secondary use is possible. Of course, the definition of property rights is deeply affected also by allocation of permitted uses. Especially in case of exclusive rights, it is very different to be entitled for a specific use, for some uses or have a residual control of choice over the usage.

[13] World Radiocommunication Conference 2015, Geneva. The concept was confirmed in the World Radiocommunication Conference 2019, held in Sharm el-Sheik.

[14] See Sims et al. (2015).

Table 7.1 A spectrum management taxonomy

Assignment \ Allocation	UNDEFINED	DEFINED
EXCLUSIVE	(IV) Residual rights to control over uses Competition but pre-emption	(III) Competition Efficient assignment but problems of adaptation and foreclosure
NON-EXCLUSIVE	(II) «free entry commons» but externalities	(I) «free entry commons» but problems of adaptation

Spectrum management, including allocation and assignment methods, can be assessed considering two main binary variables, generating a taxonomy (Table 7.1) with four possible combinations. In particular, on one side, property rights could be assigned in exclusive or non-exclusive ways and, on the other side, the allocated uses could be defined or undefined.

When assignment is non-exclusive and allocation is defined, we have a typical "free entry -commons" situation for which every operator may have access to the spectrum, but only for a very specific use (quadrant I). Here operators are generally authorised upon request by the administration, but they cannot change the use defined. This regime actually may generate problems of adaptation insofar technology evolves. In order to allow operators to adapt to technological changes in a flexible way, policy makers and regulators may allow a broader definition of allocation, i.e., allowing the same spectrum band to be deployed for different uses (quadrant II). However, once this is allowed, such flexibility may in turn generate conflicts or externalities as too many members have access to the same resource.

Here, a shift in the assignment policy, by creating a limited number of licensees with exclusive rights, may help in solving externalities problem. However, when the assignment is exclusive, the choice over the uses that licensees may perform with their spectrum bands entails other possible trade-offs. In particular, when spectrum bands are exclusively assigned, a defined allocation will mimic a fierce market competition but may hinder spontaneous adaptation to technological change and flexibility, especially if allocation is narrowly defined (quadrant III). On the other side, exclusivity with undefined allocation, or also with a broad definition of uses, (quadrant IV) would favour flexibility but reduce competition, as it may favour first-in-first-served strategies with market pre-emption purposes. When "property rights" over spectrum bands are defined and narrowly allocated, the only way to change allocation before the expiring date is to promote *refarming*, which is very costly as it generates several transaction costs. This is the reason why, in various parts of the world that have experienced refarming policies, it has been also highlighted the need for a broader allocation criterion leaving greater freedom and flexibility to licensees.

Especially in the scenario of an increased need of spectrum, rights full exclusivity might not lead to an efficient outcome. As mentioned, the sharing of resources responds to the need to maximise the intensity of their usage, beyond a mere secondary use, i.e., trading and leasing, which could have different *impasse*, due to high ex-post transaction costs and strategic behaviours. Sharing can be done using two models: (a) a non-licensed sharing that relates to non-licensed spectrum portions to be used in the "commons" perspective; and (b) a licensed sharing which refers, instead, to the spectrum used exclusively by several holders, in different geographical areas or at different times in the same area. [15]

The non-licensed sharing is a model that provides maximum flexibility and a great capacity to adapt to new uses, but clashes with the ex-ante uncertainty regarding the actual capacity available in a certain area. Moreover, non-licensed sharing, where it would prove to be effective, simply removes all entry barriers, though favours the *first comer*. Licensed sharing, instead, optimises the use of the resource from both its geographical use and the time-saturation points of view. Moreover, this approach addresses any exclusionary strategy or *impasse* in secondary contracting, which are tackled upstream in the allocation process. A hybrid system, reaping most benefits of exclusivity, mitigated by broadening the number of users. The main idea is to allow a multidimensional (e.g., time, space,

[15] Nicita and Rossi (2013).

frequency) decomposition of spectrum rights of use, as permitted by available technologies, where each licensee enjoys exclusive use of its portion of spectrum, without prejudice to incumbents' use (binary/rival use).

Each of these schemes has pros and cons that must be carefully assessed before defining any spectrum policy. Generally speaking, the case of *commons* with non-exclusive licenses should be encouraged for nomadic uses or for local business, or even for facilitating connections with low usage value in order to ensure the uninterrupted access for the customer (a typical example is the outdoor wi-fi[16]). Conversely, uses that require high investments in high added value services should remain in an area of exclusive (shared) licensing.

In any case, the choice over assignment and allocation of given portions of spectrum depends on market value, use cases, degree of competition in the upstream market and in downstream markets. The problem is often that this kind of information is private and that the asymmetry of information between policy makers or regulators on the one side, and operators on the others, may induce strategic or opportunistic behaviours by operators in the market. In particular, operators maintain strong incentives to publicly underestimate the value of spectrum, in order to induce policy makers and regulators to fix low access prices. On the other side, when low prices for spectrum access are coupled with a sufficient high number of slots, so that all the competitors may have access to a sufficient portion of spectrum, the reduced competition over spectrum prices may favour market collusion, especially in oligopolistic environments.[17]

In order to avoid opportunistic or collusive behaviours by operators, policy makers and regulators often try to overcome asymmetric information simply by letting market competition revealing the real reserve price of each operator, i.e. the maximum willingness to pay for any given portion of spectrum. In this regard, different kind of auctions, as a mechanism that mimics the market dynamics, have been suggested and implemented in many countries. Of course, we can have many different ways of designing auctions and even different ways of defining the value of the spectrum and cost of access. When a competitive price auction is chosen, under a

[16] In this direction, there is the EU WiFi4EU initiative—Free Wi-Fi for Europeans (part of the 2016 Connectivity Package), which is a voucher scheme financed by the EU for public authorities who want to offer free wi-fi connections to their citizens, for example, in and around public buildings, health centres, parks or squares.

[17] Cave and Webb (2015).

competitive bidding process, it is important to understand how bands or slots are assigned: (a) whether they are symmetric and equal, (b) how many competitors are allowed to enter, (c) whether specific pro-entrant asymmetries should be introduced, (d) what kind of conditions on territorial coverage or investments are included, (e) whether or not licensees may resale or rent spectrum and so on and so forth

However, the efficiency of market mechanisms ultimately depends on the definition of "property rights" to be assigned, as for the taxonomy described above (Table 7.1). An important lesson here comes from the pioneering works of Ronald H. Coase, who originally elaborated his famous "Coase Theorem". In his first work Coase precisely dealt with the most efficient ways to assign and allocate spectrum resources. [18] His thinking and the underlying economic concepts are still important for today's spectrum policy design.[19]

In short, the choice between administrative spectrum management (so-called *command and control*) and market mechanisms (*auctions*) strictly depends on the dimension of the compared transaction costs to be faced under the two regimes. When property rights are well-defined and transaction costs for their market exchange are negligible, market is the institution which generally performs better. These transaction costs ultimately must determine the minimum dimension of proprietary rights that allocation and assignment must define.

7.3 Spectrum Policy and the 5G Auctions in EU

The EU spectrum policy has evolved considerably over the years. An important milestone and turning point was seen in an EC study on spectrum assignment for the deployment of 5G, published in 2017.[20] The main findings showed that setting up an investment-friendly environment (through low reserve prices, long licence periods and market-driven coverage obligations) results in positive outcomes such as wider network coverage, better service quality, higher take-up and increased competition. According to the study, new approaches such as light licencing, concurrent allocations and spectrum sharing (e.g. Licensed Shared Access[21])

[18] Coase (1959). Indeed, this 1959 article anticipates many of the conclusion reached on year later about "Social Cost", Coase (1960).

[19] Hazlett et al. (2011).

[20] European Commission (2017).

[21] Licensed shared access (LSA) is a type of spectrum access that is granted to a third party by an incumbent user, typically a government body.

would allow newcomers to enter the sector more easily and new business models to take shape. These would be particularly useful in the evolving 5G environment, where network slicing could open the market to new areas not yet targeted by mobile operators.

Many of these suggestions have been incorporated within the EECC, which defines the current EU spectrum policy.[22] The Code introduces several provisions aimed at developing a systemic approach to spectrum policy and finding a balance among the different policy objectives. In particular, the EECC spectrum policy is based on: (a) an enhanced harmonisation in spectrum assignment and management; (b) a stronger support for co-investment and risk sharing; (c) the limiting of access barriers for new operators, through an increased recourse to spectrum sharing, trading and leasing; and (d) the encouragement of the use of general authorisations when possible and, for 5G deployment, light authorisations for small-area wireless access points. Furthermore, as a general prescription for radio spectrum management, Member States must ensure an effective and efficient spectrum use, pursuing the highest level of wireless broadband coverage and, in order to promote long-term investments, guarantee predictability and consistency in the granting of rights of use for radio spectrum.[23]

Under the code, granting of individual rights of use for radio spectrum must be limited to situations where such rights are necessary to maximise efficient use in demand.[24] If a Member State grants individual rights, it must be done through open, objective, transparent, non discriminatory and proportionate procedures,[25] and conditions on individual rights must be clearly established before the assignment.[26] Individual rights should be assigned for an appropriate period considering the investment amortisation, however, individual rights of use for wireless broadband services must cover a period of at least 20 years.[27] In general, undertakings may transfer or lease to other undertakings individual rights of use for radio spectrum, however, exceptions can be established by the authorities, which can also refuse the transfer if there is a clear risk that the new holder is unable to meet the original conditions.[28]

[22] See Bauer and Bohlin (2019); Lemstra et al. (2017).

[23] Art 45 EECC.

[24] Art 46 EECC.

[25] An exception to the requirement of open procedures may apply where those rights have been granted to the providers of radio or television broadcast content services.

[26] Art. 47 EECC

[27] Art 49 EECC

[28] Art 51 EECC

If Member States decide to grant rights of use for radio spectrum, NRAs must promote effective competition and avoid distortions in the internal market.[29] In doing so, regulatory bodies have several ex-ante and ex-post tools, such as (a) limiting the amount of spectrum bands for which rights of use are granted, (b) reserving a certain part of a band for new entrants, (c) attaching conditions (e.g., coverage and access conditions) to the grant of new rights of use or to the authorisation of new uses of spectrum, (d) prohibiting or imposing conditions on transfers of rights of use, and (e) amending ex-post the existing rights, to remedy a distortion in competition due to rights of use transfer or accumulation.[30]

The measures, concerning spectrum sharing and access obligations, are intended to block or limit the mentioned market distortions, such as market pre-emption and spectrum hoarding, as well as technology lock-in. Some NRAs have provided access obligation measures, intended to guarantee frequency spectrum usage for new entrants and to increase network coverage by favouring infrastructure deals and network sharing, as well as supporting third-party service provider market access.

A popular pro-competitive condition attached to individual rights is the *use-it-or-lease-it* clause, which obliges the licensee, that does not serve a specific area, to rent the frequency in order to facilitate coverage. An interesting innovative form of sharing is the so-called *club-use*, adopted by the Italian NRA AGCOM for the 26 GHz band. This regulatory mechanism allows each licensee to dynamically use all the awarded spectrum in areas where frequencies are not used by other licensees.[31] In other words, this mechanism provides that, where frequencies are not used by the other rights-holders, each licensee can use all the auctioned spectrum. Licensed operators are given incentives to sign commercial agreement to share the coverage costs, while trusted third parties can be involved with the specific task of avoiding interference and scheduling access. Moreover, awarded operators must provide access to non-telco players (service providers from vertical sectors) to develop innovative 5G services. If the required network coverage for providing a service is missing, there are NRA clauses favouring network development by third-party operators or other providers.

[29] Madden et al. (2014).

[30] Art 52 EECC.

[31] The total awarded spectrum in the Italian auction was 1GHz, allocated in 5 lots f 200MHz each.

The EECC prescribes that national authorities coordinate the use of radio spectrum, also by identifying an harmonised timeframe where the use of specific radio spectrum is to be authorised. The EECC focuses on 5G, challenging so-called pioneer bands to be made available to operators by 2020[32] (see Box 7.1).

Box 7.1 5G Auctions in Europe

The three 5G pioneer radio spectrum bands are the 700MHz, harmonised through an EC Implementing Decision[33] in 2016, the 3.6GHz (3.4-3.8GHz)[34] and the 26GHz (24,25-27.5GHz) frequency ranges.[35] The EECC requires MSs to make them available to operators by the end of 2020, taking all appropriate measures to reorganise and allow the use of sufficiently large blocks of the 3,4-3,8 GHz band and allow the use of at least 1 GHz of the 24.25-27.5 GHz band.

Where the assignment is concerned, as for end 2019, the procedures have been completed or are ongoing in 12 Member States. In terms of numbers, the assignment regarded 23.2% of 700MHz spectrum, 38.3% of 3.4-3.8GHz spectrum and 3.6% of 26GHz spectrum. A synthesis (at the end of 2020) is provided by the DESI sub-index referring to *5G readiness* (see Fig. 7.1 below).[36]

For 26GHz, Italy is the only country which has assigned spectrum portions in this band (and in all three mentioned[37]). Finland is

(*continued*)

[32] Art. 53 and 54 EECC.

[33] Commission Implementing Decision (EU) 2016/687 of 28 April 2016 on the harmonisation of the 694-790 MHz frequency band.

[34] Commission Implementing Decision (EU) 2019/235 of 24 January 2019 on amending Decision 2008/411/EC as regards an update of relevant technical conditions applicable to the 3400-3800 MHz frequency band.

[35] Commission Implementing Decision (EU) 2019/784 of 14 May 2019 on harmonisation of the 24,25-27,5 GHz frequency band for terrestrial systems capable of providing wireless broadband electronic communications services in the EU.

[36] European Commission (2019).

[37] As stated above, MSs have adopted a common deadline for the effective usability in the 3.6 GHz band and 26 GHz band, to be assigned by the end of 2020.

Box 7.1 (continued)

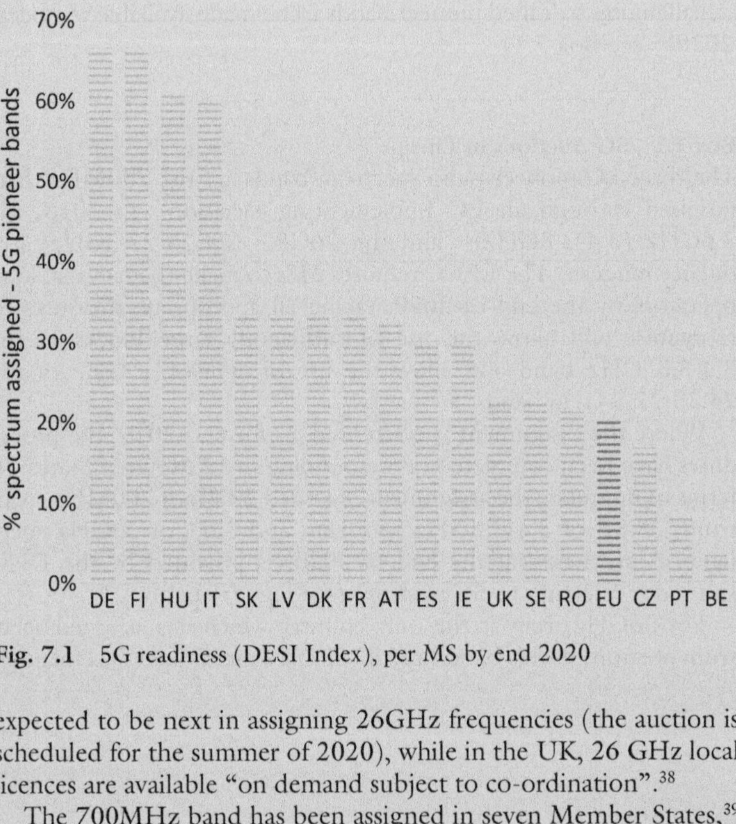

Fig. 7.1 5G readiness (DESI Index), per MS by end 2020

expected to be next in assigning 26GHz frequencies (the auction is scheduled for the summer of 2020), while in the UK, 26 GHz local licences are available "on demand subject to co-ordination".[38]

The 700MHz band has been assigned in seven Member States,[39] while 12 countries[40]have assigned the 3.6 GHz. 26 countries plan to

(*continued*)

[38] The technical conditions for "Shared Access Indoor 26 GHz" were published by Ofcom in the document "Draft UK Interface Requirement (IR) 2105" (July 2019).

[39] Those countries are Germany (2015), France (2015), Finland (2016), Italy (2018), Sweden (2018), Denmark (2019) and Hungary (2020).

[40] Countries which assigned 3.4-3.6 GHz bands are: Austria, Czech Republic, Finland, Germany, Hungary, Ireland, Italy, Latvia, Slovakia, Spain, Romania, and UK.

Box 7.1 (continued)

hold auctions for at least one of the three bands during 2020. Unfortunately, the outbreak of COVID-19 has delayed auctions in Austria, France, Spain and Portugal. The Czech Republic tender was postponed for different reasons, while, in Hungary, the auction took place just before the restrictions, assigning frequencies in the 700MHz, 2100MHz and 3.6 GHz band portions.[41]

In terms of final award prices, auctions in Italy and Germany were the most competitive (raising €6,550 billion each). The other countries where auction total outcome was higher than 1 billion were the UK (€1.55 billion) and Spain (€1.41 billion), and the total investments for 5G spectrum licenses in the EU 27+ amounted to about €17 billion. A popular way to compare spectrum award prices, for the 3.4-3.8 GHz bands[42] , is to normalise final award prices for MHz auctioned, population covered, and licence duration. Under this approach, Italian operators are those which paid the most, about €20/MHz per inhabitant for each licensed year. In Germany, that amount, normalised in the same way, reached €8.39/MHz, while the same value stood at €7.55 in Spain and at €5.73 in the UK.

It is however important to note that this benchmark approach (often used to evaluate how much pro-market or pro-state finance an auction is) does not consider that: (a) what is primarily relevant for identifying a "revenue maximising" objective is the reserve price, while the final award price depends on the competitive bids of rational participants, which assess its value and the possible return on their investment;[43] (b) it is true that the competitive dynamics within an auction can be modulated also by adjusting lots, i.e., few lots comprising more spectrum intensify competition and increase final price, however, the definition of the lots size has primarily a technical objective, since the quantity of contiguous spectrum is important to the quality of 5G

(*continued*)

[41] No bids were submitted for the 26 GHz spectrum.

[42] According to the auction already held, a direct comparison is only possible for the 3.4-3.8 GHz band.

[43] For example, in the Italian auction the overall reserve price was 2.5bn€ and the final award price was 6.5bn€

Box 7.1 (continued)

services that can be provided;[44] (c) this in turns implies that the value of 1 MHz of spectrum also depends on the size of the lot adjudicated, due to (c_i) the different quality of services that can be provided and, (c_{ii}) the competitive advantage that an operator adjudicating a large amount of spectrum (which is a scarce resource) acquires over other operators.

Of course, the competitive advantage and the overall value of the spectrum frequency adjudicated also depends on the conditions attached to the individual rights of use, i.e., coverage and access obligations. As for coverage obligations, spectrum assignment involved obligations in 9 countries. Obligations for the 700MHz band were set up in Finland, France, Germany, Italy, the Netherlands, and Sweden. Regarding the 3.4-3.8 GHz band, coverage obligations were established in the Czech Republic, Germany, Italy and Poland. There are no coverage obligations for the 26 GHz band, both in the Italian licensed frequencies and in the French 5G Plan for the same band. Coverage obligations are often different according to the amount of MHz awarded, and between established operators and new entrants, to whom a portion of spectrum is often reserved. Regarding access obligations, a variety of tools have been adopted by most countries. For example, in Italy, in the 700 MHz band, awarded operators must allow roaming access to new entrants, however, they can only benefit from this right by covering at least 10% of the population with their own frequencies. In the 3.6-3.8 GHz band, operators holding at least 80 MHz must provide access to other non-licensee players. The UK has set up a variety of policies for different bands by introducing spectrum sharing in 3.8-4.2 GHz, 1800 MHz and 2300 MHz and in the 26 GHz band (only for indoor services). In Austria, to balance the competition, the NRA can grant newcomers access to an existing operator's network for a limited period. In Finland, some portions of the 26GHz band are available for shared use.

[44] For example, "5G needs a significant amount of new harmonised mobile spectrum so defragmenting and clearing prime bands should be prioritised. Regulators should aim to make available 80-100 MHz of contiguous spectrum per operator in prime 5G mid-bands (e.g. 3.5 GHz) and around 1 GHz per operator in high-bands (e.g. mm Wave spectrum)." GSMA (2020).

As mentioned above, 5G, especially thanks to network slicing, is resulting in technology-shift paradigms, some of which cannot be as yet forecasted. This gives rise to new services and business models, where traditional telecom players interact in different ways in an extended value chain, including third-party operators, such as those from the *vertical industries* (automotive, energy, healthcare, etc.). This ecosystem requires a maximum market flexibility and openness. However, it is crucial, at the same time, that the flexibility does not hamper predictability and consistency in the granting of rights of use for radio spectrum and, therefore, overall investments.

The Code aims to guarantee a balance in creating a framework which is effectively competitive and open to innovation. However, this could be negatively affected by an inconsistent implementation at national level. Therefore, harmonisation and coordination mechanisms are even more important for spectrum management than for the general consistent application of rules in the internal market (see Sect. 2.4). Indeed, this is an essential part of an efficient and effective systemic spectrum policy approach.

References

Bauer, J., & Bohlin, E. (2019). The Role of Regulation in 5G Market Design. Quello Center Working Paper.

Cave, M., & Nicholls, R. (2017). The Use of Spectrum Auctions to Attain Multiple Objectives: Policy Implications. *Telecommunications Policy, 41*(5-6), 367–378.

Cave, M., & Webb, W. (2015). Spectrum Management Using the Airwaves for Maximum Social and Economic Benefit, CUP.

Coase, R. (1959). The Federal Communications Commission. *Journal of Law and Economics, 2*, 1–40.

Coase, R. (1960). The Problem of Social Cost. *Journal of Law and Economics, 3*, 1–44.

Ericsson. (2019). Mobility Report.

European Commission. (2016). Communication on 5G for Europe: An Action Plan—COM(2016) 588 final.

European Commission. (2017). Study on Spectrum Assignment in the European Union.

European Commission. (2019). Digital Economy and Society Index (DESI)—Connectivity Report.

GSMA. (2019). The Mobile Economy.

GSMA. (2020). 5G Spectrum: GSMA Public Policy Position.
Hazlett, T. W., Porter, D., & Smith, V. (2011). Radio Spectrum and the Disruptive Clarity of Ronald Coase. *Journal of Law and Economics, 54*(S4), 125–165.
Lemstra, W., Cave, M., & Bourreau, M. (2017). Towards the Successful Deployment of 5G in Europe: What Are the Necessary Policy and Regulatory Conditions? CERRE report.
Madden, G., Bohlin, E., Tran, T., et al. (2014). Spectrum Licensing, Policy Instruments and Market Entry. *Review of Industrial Organisation, 44*, 277–298.
Nicita, A., & Rossi, M. A. (2013). Spectrum Crunch vs. Spectrum Sharing: Exploring the 'Authorised Shared Access' Model. *Communications & Strategies, 90*, 17–40.
Pogorel, G. (2018). Spectrum 5.0 Rethinking Spectrum Awards for Optimal 5G Deployment, presented at the ParisTech Conference: Spectrum 5.0.
Sims, M., Youell, T., & Womersley, R. (2015). *Understanding Spectrum Liberalisation*. Boca Raton: CRC Press.

CHAPTER 8

Digital Transformation and Electronic Communications

Abstract Digitalisation has been deeply impacting all economic sectors; nevertheless, telecom as well as media markets are at the forefront. Digitalisation implies a process of multimedia convergence allowing for a range of different digital content and services to be transmitted on the same digital network. Moreover, the decoupling of networks and services has resulted in an ecosystem modularity which supported and provided incentives for the emergence of global on-line platforms, offering content and services over-the-top (OTT) of the traditional telecom value chain. These factors have dramatically influenced market dynamics in telecom and media markets, drawing a great deal of attention from the public, policymakers and stakeholders. Crucial topics concerning the relationship between OTTs and traditional players, such as a regulatory playing field and network neutrality, have recently been at the core of policy debate and will continue to fuel it in the coming years.

Keywords Digital transformation • Digital platforms • OTT service • Net neutrality • Multimedia convergence

A. Manganelli, A. Nicita, *The Governance of Telecom Markets*, Palgrave Studies in Institutions, Economics and Law, https://doi.org/10.1007/978-3-030-58160-2_8

8.1 Digitalisation and Multimedia Convergence

In the previous chapters, we have analysed the composite evolution of telecom regulation and markets in Europe over twenty years, under both a legal and economic perspective. A fundamental dynamic, further exacerbating complexity, is the impact of digitalisation and digital technologies on communication services. This impact has been enormous and is increasingly affecting patterns of consumption, business models, definition of relevant markets and the overall vertical distribution of value along the supply chain.

As previously mentioned, since the early phase of liberalisation, technological innovation and digital transformation have exponentially decreased telecommunication network deployment costs and increased telecom service cost-efficiency, thus, changing the economics of the sector and making it profitable for alternative operators to build alternative networks or alternative network segments.

On the other hand, digitalisation has strongly influenced markets and services by means of (a) digitisation of information and signals, (b) development of transmission techniques on the Internet protocol, and (c) the development of multimedia reception devices. One of the major consequences has been the transformation of "special purpose" service delivery networks, dedicated to a specific service e.g., voice communications or cable television, to "general-purpose" electronic communications networks capable of supporting multiple applications and services including connectivity and transmission.[1]

The decoupling of service from transmission, jointly with a principle of *technological neutrality* of transmission networks, have been reckoned within the European regulation. Since 2002, there has been in place a single legal framework for all transmission networks in which electronic communication networks are generally defined as transmission systems permitting the conveyance of signals, irrespective of the technology adopted and the type of information conveyed.[2] The following are all considered electronic communications networks: (i) satellite networks, (ii) fixed (circuit- and packet-switched, including Internet) and mobile terrestrial networks, (iii) electricity cable systems, to the extent that they are used for the purpose of transmitting signals, (iv) networks used for radio and television broadcasting, and (v) cable television networks.

[1] OECD (2016).
[2] Art. 2(1) EECC.

Even if all the networks are subject to the same regulatory framework, services delivered over those networks (even over the same network) are disciplined by different regulatory domains. There are:

(a) electronic communication services disciplined by the EECC,[3] which encompass (1) "internet access service";[4] (2) interpersonal communications services;[5] and (3) services consisting wholly or mainly in the conveyance of signals such as transmission services used for the provision of machine-to-machine services and for broadcasting, with the exception of services exercising editorial control over content;
(b) media content regulated by the Audio-visual Media Directive (AVMS), which governs EU-wide coordination of national legislation on all audio-visual media content over: (1) traditional TV broadcasts, (2) on-demand services, and on some aspects of (3) video-sharing services;[6]
(c) information society services, involving "any service normally provided for remuneration, at a distance, by electronic means and at the individual request of a recipient of services", whose rules are provided by the E-Commerce Directive.[7]

[3] Art. 1(4–7) EECC.

[4] "Internet access service" means a publicly available electronic communications service that provides access to the Internet and, thereby, connectivity to virtually all end points of the Internet, irrespective of the network technology and terminal equipment used. Point (2) of the second paragraph of Article 2 of Regulation (EU) 2015/2120—TSM regulation.

[5] "Interpersonal communications service" means a service normally provided for remuneration that enables direct interpersonal and interactive exchange of information via electronic communications networks involving a finite number of persons, whereby the persons initiating or participating in the communication determine its recipient(s) and does not include services which enable interpersonal and interactive communication merely as a minor ancillary feature that is intrinsically linked to another service; Art. 1(5) EECC.

[6] Directive 2018/1808. An audio-visual media service (AVMS) has the principal purpose to provide programmes, under an editorial responsibility of a media service provider to the general public, in order to inform, entertain or educate, by means of electronic communications networks. An AVMS could be either (a) a television broadcast. i.e., for simultaneous viewing of programmes on the basis of a programme schedule (linear AVMS) or (b) an on-demand AVMS, for the viewing of programmes at the moment chosen by the user and at his/her individual request on the basis of a catalogue of programmes selected by the media service provider (non-linear). A video-sharing platform service has the principal purpose to provide programme or user-generated by means of electronic communications, for which the platform does not have editorial responsibility.

[7] Directive (EC) 2000/31, on certain legal aspects of Information Society Services, in particular electronic commerce, in the Internal Market, and Directive (EU) 2015/1535 laying down a procedure for the provision of information in the field of technical regulations and of rules on Information Society Services (codification).

This distinction regarding the three types of services and related bodies of rules was quite relevant when it was originally introduced. However, nowadays, it does not seem to address the current market dynamics, i.e., substitutability between digital services and the overlapping of those categories (especially some audio-visual media contents and information society).[8] Migration to broadband all-IP networks has certainly decoupled services from transmission, making it possible for service, applications, and content to flow seamlessly across networks, thus, fundamentally changing the markets.

The first effect was to push the telecom industry, as well as media and information technology sectors towards a *multimedia convergence*, allowing both traditional and new communication services—whether voice, data, sound or video—to be provided simultaneously over many different networks and often received with the same multimedia device.[9] As a consequence, telecom operators started to commercialise a much broader variety of services. The provision of fixed and mobile communication, audio-visual media services, i.e., *multi-play offers*, are today widespread, along with other innovative value-added information society services developed at the edge of the communication systems (See Box 8.1).

Box 8.1 Multi-play bundled offers

Multimedia convergence has led telecom companies to provide multiple services, i.e., *multi-play offers*.[10] Multi-play offers, including both fixed broadband and television, currently represent more than half of total broadband subscriptions in Europe.[11]

Multi-play offers are bundles of services that are jointly provided by the same provider under the same or linked contract at a joint price, usually discounted compared to the stand-alone services and product prices. This commercial practice is called *pure bundling* and, in the case these services or products can also be purchased separately, it is called *mixed bundling*.

(*continued*)

[8] De Streel and Larouche (2016).

[9] European Commission (1997).

[10] With a focus on television, triple-play services refer to bundles of fixed voice, broadband and pay-television services, and quadruple-play services to triple-play bundles plus mobile services, voice and/or data.

[11] BEREC (2018).

Box 8.1 (continued)

Bundling practices often bring benefits to consumers by (a) reducing transaction costs related to the contracting and dealing with multiple providers, (b) allowing pricing discounts reflecting the economies of scope that companies can develop in a convergent environment, e.g., fixed costs such as network costs or customer care that can be shared across multiple services, and (c) allows market players to promote new services, benefiting the consumers where elements of the bundle are complements.[12]

However, bundling can raise concerns in terms of competition, as bundling could be a way to leverage market power from a dominated market to a competitive one, e.g., a dominant cable or satellite TV services provider offering pure bundling with broadband services. As for consumers, a bundling practice may be detrimental as it can make switching more difficult or costly and raise the risks of *contractual lock-in*. Moreover, bundles may also reduce consumer choices because pure bundling may include one or more services which the consumers may not value, e.g., BB services bundled with fixed telephony where end-users only use mobile voice services.

Bundled offers have become increasingly widespread and are an important element of competition (see Fig. 8.1).

For this reason, the EECC seeks to attenuate the pure bundling approach with the aim of fostering competition and empowering consumers. Consumer protection and empowerment rules such as termination and duration of contracts and switching practices (as described in Sect. 4.2), that are applicable to a given service when provided as a stand-alone service, must also be applicable when that service is part of a bundle. In general, any contractual issue must be governed by the rules applicable to the respective stand-alone element of the bundle, however, in order to further support consumers, a right to terminate the contract related to an element of a bundle because of a lack of conformity or a failure to supply, gives a consumer the right to terminate all elements of the bundle.[13]

(*continued*)

[12] OECD (2015).

[13] Recital 283 and art. 107—Bundled Offers—of the EECC.

Box 8.1 (continued)

Fig. 8.1 Multi-play subscriptions in 2017, % households (Data from European Commission (2018))

Moreover, video and information society service and applications, on the one hand, and networks, on the other, have taken a co-evolutionary path. The former has been supported by enhanced internet network transmission capacity, allowing an ever-increasing mass of data to be delivered (see Chaps. 6 and 7). On the other hand, the constantly growing spread of fixed and mobile internet connections along with the increasing number of users interconnected "anytime, anywhere" has been sustained by the availability of consumer-attracting or even *must-have* services, applications and contents.

In Fig. 8.2 below, we can appreciate the evolution of fixed internet connections, their increasing speed and internet users, as well as the impressively growing percentage of people accessing the Internet through mobile data connections.

8.2 Digital Platforms and the Level Playing Field

The multimedia convergence and modularity of broadband all-IP systems, apart from encouraging telecom multi-play offers, has enabled new native digital players to penetrate the markets. These new market actors have introduced a wide range of digital products and services, which are either substitutes or complements to those of the telecom and media companies.

These include Voice over IP communications, instant messaging, video-on-demand, video-sharing, online advertising as well as e-commerce, marketplace services, internet search engines, social media, products price comparison, application distribution, payment system services, collaborative activities, etc.[14]

These digital players are called Over-The-Top (OTT) providers as they offer services to end-users via the telecom infrastructures *over the top* of the traditional telecom market value-chain, extending and transforming it into a non-linear digital ecosystem.

The main OTTs compete in the global markets, interconnecting an enormous, and growing, number of users, and interacting with many diverse companies, such as traditional communications and media players, advertisers, applications producers, vertical companies, and small specialised companies.[15] The result is a networked modular ecosystem, permeated by complex business relationships and pervasive user networks made up of economic, social and operational (human and non-human[16]) interactions. The digital ecosystem's fundamental technological backbone is made up of fixed and mobile electronic communications physical infrastructures that enable digital services, content and applications. These services at the edge of the ecosystem are increasingly becoming the system's commercial and economic core (for more information about OTTs and the Big Techs, see Box 8.2).

Box 8.2 Big Techs, two-sided platforms and digital data

The main and most popular OTTs are few global companies, i.e., Amazon, Facebook, Netflix, Google, Apple and Microsoft which are patchily grouped, and also known as GAFA, GAFAM or FAANG,[17] or more generally as *Big Techs*.[18] All of them, except for Netflix, were

(continued)

[14] See European Commission (2016). See also BEREC (2016).

[15] See European Commission (2016).

[16] As previously described, Sect. 3.3.1, non-human interactions among objects connected, often through dedicated platforms, are increasingly important in the digital ecosystem.

[17] GAFA stands for Google, Apple, Facebook, Amazon; while GAFAM adds Microsoft to the group; and FAANG reshuffles the order and substitutes Microsoft with Netflix.

[18] The Big Techs have somehow replaced the so-called "Big Oil", the large multinational energy companies (Exxon Mobil, BP, Gazprom, PetroChina and Royal Dutch Shell) in the collective capitalistic imaginary, and also from the top of the USA NASDAQ stock index.

Box 8.2 (continued)

among the world's top 10 leading companies in 2019 in terms of market capitalisation, along with the Chinese digital platforms Alibaba and Tencent. Microsoft is the leading company globally (see Fig. 8.2). Concerning revenues, Big Tech average income was €115 bn in 2018, up 35% compared to 2016, against the main world telecom and media companies' average of less than €30 bn, up 3% from 2016. The top income was reported by Apple (€225 bn, +15%), followed by Amazon (€197 bn, +61%) and Google (€116 bn, +42%). All these figures reflect the clear and growing prominence of OTTs in the twenty-first century's ***digital capitalism***.[19]

Big Techs have different features. Besides performing different activities, one of the main distinctions concerns revenue sources and

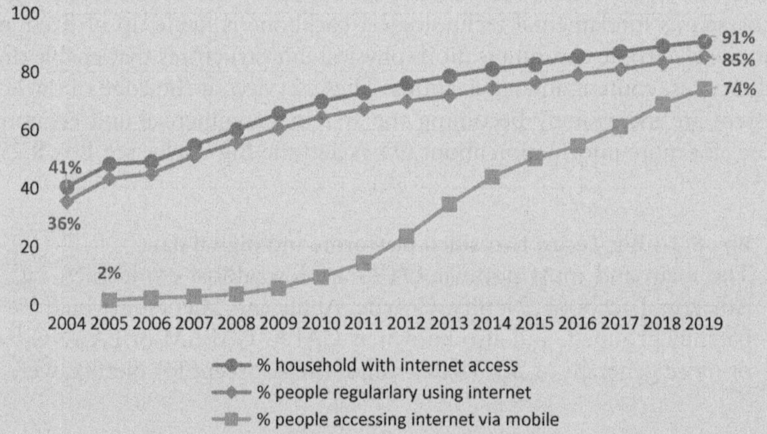

Fig. 8.2 Fixed and mobile internet development in EU (Data from European Commission (2019))

(continued)

[19] Schiller (2000).

Box 8.2 (continued)

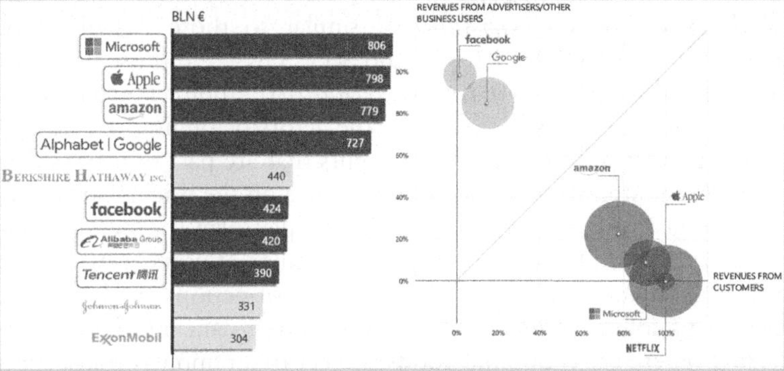

Fig. 8.3 Big Techs market capitalisation and source of revenues (AGCOM 2019)

the associated business model. As shown in Fig. 8.3: (a) Netflix, Microsoft and Apple's incomes are based on selling services and products to end-users; (b) Google and Facebook's revenues come primarily from advertisements; and (c) Amazon has a more hybrid business model, even if revenues from customers are predominant.

Despite these differences, all Big Techs work as *digital platforms* in the sense of the multi-sided market theory,[20] i.e., they operate in markets with two (or more) sides characterised by two (or more) different and separate groups of economic agents. Interactions among these groups are mediated by platforms, which work as *matchmakers* drastically reducing transaction costs, and taking into account that demand from at least one group of customers depends on demand from another group of customers (so that these are not externalities for the platform). Customers of the two groups do not take these indirect network effects into account (so that they are in fact exter-

(continued)

[20] See, for example, Rochet and Tirole (2006); Armstrong (2006); Filistrucchi et al. (2013).

Box 8.2 (continued)

nalities for buyers). In these markets, there are both the classic network effects (*direct externalities*)—similar to those described for telecom companies (see Chap. 1 and Sect. 5.3)—and those typical of a multi-sided market (*indirect externalities*). Indirect externalities occur when the decisions made by the actors belonging to one side of the market produce effects on agents that are part of other sides, and whose intensity has a decisive influence on the pricing structure.

The basic common element of digital platform business models is *digital data*. Due to the incredible number of users, as well as to the time spent and to the high number of actions made online, platforms collect an incredible amount of user data and *digital footprints*. These are economic goods subject to an implicit transaction against free or cheaper services.[21] This market mechanism allows platform *algorithms*, based on *deep learning* and *artificial intelligence* tools, to extract valuable information and to profile individual demand for services and products for all and each user, in terms of preferences, needs and willingness to pay. Platforms rely on profiling to monetise data by either using information directly, as retailers for online trade, or indirectly, mainly as recipients of targeted personalised advertising. (Fig. 8.4) Moreover, data allows algorithms to progressively improve themselves as new data is analysed, by improving estimations on user preferences and, thus, on profiling. Platforms are then able to increase their matching capacity and, accordingly, increase demand on both sides of the platform. On the user side, by increasing service productive efficiency and quality (i.e. services or products are suited to user preferences and needs); on the other group side (e.g., advertisers), by increasing the number or attention of users to whom ads are addressed. All this produces incentives for platforms to collect more and more data by adopting *zero price* business models (i.e., no monetary payment from end-users) and to expand scale and scope of their activities

(*continued*)

[21] Compared to a theoretical counterfactual scenario with no data implicitly exchanged and monetised.

Box 8.2 (continued)

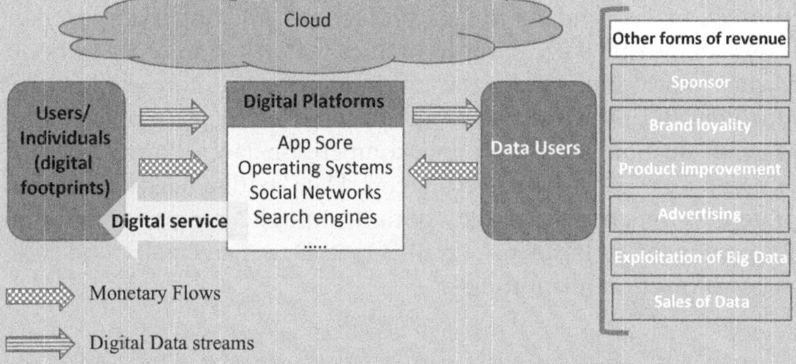

Fig. 8.4 Simplified representation of digital platforms two-sided market (Agcom 2018)

and, overall, capturing a significant portion of the digital ecosystem market value. In turn, big online platform providers are leveraging their gatekeeper role to entrench and extend their market power. Consequently, their market position is becoming increasingly more difficult to contest by potential competitors.

Notably, the market power enjoyed by large platform providers translates into a "data advantage" whereby these dominant players can set unilateral rules and autonomously foreclose access and use of data. Similarly, it is likely that similar imbalances would concern access to co-generated IoT data from consumer and industrial devices in the near future. The European Commission acknowledges that today "a small number of Big Tech firms holds a large part of the world's data."[22] The EC intends to design a third-way *Data Strategy* that avoids concentration effects (experienced by the privately led US economy) while preserving high privacy, security, safety and ethical standards (thereby marking a clear difference from the Chinese model). A European Data Space will be at the centre of the EU data strategy for the next five years, characterised by smooth

(continued)

[22] European Commission (2020).

Box 8.2 (continued)

forms of data sharing among consumers and businesses, effective implementation of EU rules (in particular, competition law, data protection law and consumer protection) and international data flows complying with European values.[23] Further, in order to boost technological innovation, the EC intends to promote data pools[24] enabling Big Data[25] analytics and machine learning, compliant with data protection legislation and competition law, allowing for the emergence of data-driven ecosystems. Data sharing needs to be facilitated as much as possible, both for public sector information (government-to-business—G2B—Open Data Directive)[26] and business-to-business relationships.

OTTs' innovation at the edge of the digital ecosystem is progressively turning into a ***disruptive innovation*** for traditional well-entrenched telecom and media companies. This is due to a paramount impact on internet traffic, network efficiency and dimensioning (see next Sect. 8.3). Moreover, some OTT services could well be substitutes to telecom and media company services, consequently, competing for the same end-users. This is the case, for example, for VoIP interpersonal communications, and pay audiovisual media services, or for the same intermediation role between two market sides, i.e., user attention and advertisers.[27] Therefore, there is an issue concerning the *regulatory playing field*: different, much lighter, regulatory obligations placed on platforms, resulting in a competitive disadvantage for telecom and media companies.

[23] European Commission (2020).

[24] Data pools would act as a one-stop shop allowing entities interested in having access to the data to interact with a single administrator. Companies could send their data to a platform and get back aggregate data with no indication of which company it comes from.

[25] Big Data can be defined as a collection of information data so extensive in terms of volume, speed and variety to require specific analytical technologies and methods for the extraction of value or knowledge. See Delmastro and Nicita (2019); Stucke (2016) Big Data and Competition Policy.

[26] Directive (EU) 2019/1024, repealing Directive 2003/98/EC as revised by Directive 2013/37/EU.

[27] See BEREC (2016); OECD (2014). In this regard, the EECC requires that NRAs must have an adaquate "regultory capacity" in the meaning that they must be equipped by Member States with "adequate financial and human resources to carry out the task assigned to them". Art. 9 (2) EECC. This general provision concerns only organisational resources, which are

This regulatory imbalance can be explained by very different reasons: (a) the existing (yet evolving) regulatory distinction among electronic communications, audio-visual and information society services which may result in framing competing services into different regulatory regimes; (b) an effective appreciation of the different economic, business and technical aspects of digital undertakings and consumers, requiring a different regulation; (c) a deliberate policy approach leaving digital markets unregulated or regulated with a ***light-handed approach***, at least in the short term; (d) a limited ***regulatory capacity***[28] suffered by public bodies, not allowing regulators (namely, telecoms, media and data protection) and competition authorities to design an effective public intervention.

A light-touch policy approach for digital markets (c) has been so far preferred because: (c_i) the incomplete information about innovative markets increases the risk of a regulatory intervention that might hinder innovation; however this risk has been progressively part of a regulatory trade-off involving on the other side the risk that unregulated market dynamics could lead to direct or indirect consumer harm; (c_{ii}) a significant part of the public opinion has been seeing the Internet and the web as a place of individual freedom to be preserved, under all circumstances, and, therefore, a locus not to be subject to (external) rules.[29] However, this ***romantic*** vision of the Internet as a place of individual freedom does not seem to adequately reflect the changes occurred on the web, shifting from an almost atomistic decentralisation of actors to a strong centralisation involving only a few global platforms.

The limited regulatory capacity (d) hinges upon two main elements related to the relationship between public bodies and digital platforms: (d_i) a marked information asymmetry, due to the continuous and fast digital technological innovation that causes public bodies to always lag behind in terms of understanding digital market dynamics and platform internal technical functioning (e.g., the platform algorithms as ***black boxes***); (d_{ii})

extremely important, however one of the main regulatory capacity deficit is currently related to NRAs' regulatory powers and competences, which are defined by the overall legislation.

[28] OECD (2009); BEREC (2013).

[29] John Perry Barlow, A Declaration of the Independence of Cyberspace, 8 February 1996, "Governments of the Industrial World, … you have no moral right to rule us nor do you possess any methods of enforcement we have true reason to fear. … You do not know our culture, our ethics, or the unwritten codes that already provide our society more order than could be obtained by any of your impositions. … We are forming our own Social Contract. This governance will arise according to the conditions of our world, not yours. Our world is different."

the institutional public policy setting, which is traditionally designed and implemented nationally, while digital and platform economies are global in nature, and based on their virtual context often has managed to elude public regulatory entanglements.

In this regard, the EU has been taking steps on all the points mentioned above, and has progressively included digital platforms and their services within the scope of the directives governing the communications and media services at supranational level. The main adaptations of EU rules to the digital convergence has taken place in the audio-visual sector, which has seen an earlier and more substantial transformation due to digital services compared to other sectors.

In 2010, the first AVMSD directive was issued to respond to the convergence deriving from the transmission of multimedia content on an increasing number of platforms. The AVMSD amended the Television Without Frontiers Directive, which covered only traditional broadcasting services (linear services), extending its scope to include video-on-demand (non-linear) services with an editorial nature. However, only a few rules were extended symmetrically, while most involved a *lighter touch*, due to the higher degree of consumer control in on-demand services. The recent amendments in 2018 further adapted the AVMSD to digital transformation, imposing a greater number of symmetric rules for on-demand services. Moreover, it also required video sharing platforms (previously completely excluded from the discipline) to adopt a few of the social regulation measures for the protection of minors, control of hate speech and compliance with the qualitative rules on advertising.

On the electronic communications front, the 2018 EECC introduced some regulatory measures regarding OTT platforms for the first time, categorising OTT communication services as electronic communication services (ECS). The EECC adopted "a future-oriented definition of electronic communications services … build on a functional approach … [as] from an end-user's perspective it is not relevant whether a provider conveys signals itself or whether the communication is delivered via an internet access service."[30]

In particular the EECC divided the interpersonal communication services (ICS) into two sub-categories: (a) number-based ICS services, corresponding to the traditional fixed and mobile voice services, in which the service is connected with numbers in numbering plans,[31] assigned by

[30] Recital 15 EECC.

[31] *Numbers* are scarce resources, like spectrum, and the EECC defines conditions for individual assignments of rights of use and efficient use, and management through the definition of a numbering plan (art. 93 and 94 EECC).

public authorities for the routing of traffic, and not only as a user identification, and (b) the ICS services independent of the number, provided by the OTTs (e.g., Skype, WhatsApp) where the number is only the user's identification and not assigned and used for routing operations. Number-independent services are subject to a light-touch regulation, e.g., OTTs are explicitly exempted from the general authorisation regime, as they do not benefit from the use of public numbering resources, and do not participate in a publicly assured interoperable ecosystem. However, an interoperability obligation for number-independent ICS reaching a significant level of coverage and user uptake is foreseen, in circumstances where end-to-end connectivity between end-users would be endangered.[32] This provision is subject to a vertical coordination mechanism, requiring a Commission decision outlining what measures can be taken by NRAs. OTTs are also exempted from most end-user rights regulations, with some exceptions such as the provision on contract information,[33] and the inclusion in comparison tools, when applicable, that is, when number-independent interpersonal communications services are on payment.

8.3 Net Neutrality Debate(s) and the EU Open Internet Regulation

The issue of a level playing field between telecom and media companies, on the one hand, and OTTs, on the other, is based on the substitutability of some services, either toward end-users or other groups in the multi-sided markets, e.g., advertisers. As previously mentioned, this phenomenon is principally related to media companies and bundles of services offered by telecom companies.

However, services provided by telecom companies and OTTs have mainly complementary features, the former providing internet access services, and the latter providing content and applications to end-users. To short, an end-user needs both a broadband connection and online content and applications. Complementarity of inputs is one of the main features of the digital ecosystem and its modular functionalities.

[32] Art. 61 (2) c EECC.
[33] Art 102 EECC.

Therewith, this complementarity is asymmetric as Internet Service Providers (ISPs) have the material capacity to inspect and manage data traffic (i.e., *throttling*, prioritising or blocking), consequently, affecting the service quality (or even the provision) of content and application providers (CAPs), while the opposite is not possible. In other words, ISPs can apply non-price discrimination practices on CAPs' services. Moreover, depending on their scale and the competitive environment, ISPs can be (competitive) bottlenecks for CAPs to access end-users and, thus, may exert market power, if the latter has no countervailing power.

This situation was a no brainer when internet usage mainly involved simple email exchanges and web browsing, with a quite decentralised and symmetric data flow, needing low bandwidth. Today, instead, almost 95% of internet data packets are related to Voice Over IP, video talk, video streaming and multi-player gaming, which are mostly unidirectional—from content providers to users—and use high bandwidth with a quality of service that decreases drastically with delays. This dramatic growth of bandwidth requirement was costly for ISPs. On the contrary, it was linked to a huge revenue source for CAPs, mainly from end-users and/or advertisers, without contributing to the ISPs' network fixed costs.[34]

In this context, ISPs policy arguments and market strategies focused on (a) the importance of efficient traffic management practices to make efficient use of networks, namely prioritising services negatively affected by delay and avoid congestion, and (b) the need to charge and discriminate on prices and quality of service for CAPs, in order to incentivise network investments and increase bandwidth quality and capacity, thus internalising the existing spill-over effects on CAPs.

This market transformation and the ISPs strategies underlie the *network neutrality* debate, whose original conceptualisation was expressed as "all data packets on the Internet should be treated equally."[35] Following, the net neutrality concept took on many different hues and forms depending on the market and policy context. Its consolidated basic definition concerns: (a) the prohibition of payments from content providers to internet service providers; and (b) the prohibition of traffic prioritisation (*fast lane*

[34] It is important to note that OTTs pay transit fees or peers from their networks with infrastructure owners and, of course, end-user subscribers pay for service providers for the bandwidth.

[35] Wu (2003).

versus *dirty roads* or *managed services* versus *best effort*), with or without compensation.[36]

As a consequence, net neutrality is about discrimination and differentiation of treatment by ISPs with anticompetitive effects on CAPs, which may be technically or economically pushed out of the market or obliged to downgrade the quality of services to end-users, who will see a reduction in the benefits they receive from their internet connection. This argument was mainly developed in the US debate, where dominant and vertically integrated ISPs (e.g. Comcast, AT&T) offer in bundles communications and audio-visual managed services competing with OTT operators. This situation provides ISPs with incentives and opportunities to adopt anti-competitive behaviour (e.g., exclusion, excessive pricing).[37]

Later, also in EU countries, triple and quadruple-play bundles including digital television started to become widespread. However, no EU operators have developed a vertical integration between internet access and audio-visual content as in the US—where Comcast merged with NBC-Universal in 2011 and with Sky in 2018, while AT&T merged with Time Warner in 2018. Moreover, differently from the US, where the broadband market is very concentrated end-to-end, most EU end-users can choose, also due to the LLU, from a number of (more or less) infrastructure-based internet access providers (see Sects. 2.2 and 5.2). Finally, as described in Sect. 3.2, in Europe a quite different institutional setting exists regarding competition law and regulations. The fact that US Supreme Court's consolidated case law (i.e., the *Trinko* decision) restricts antitrust enforcement in regulated sectors made it more compelling to develop a regulatory definition of net neutrality aimed at preventing anti-competitive effects.

Without any net neutrality obligations, ISPs would set a system of data paid termination, as for voice calls. Anti-competitive concerns would arise that an ISP would charge (high) differentiated termination fees, combined with non-price discriminations, to reduce competition and exclude competing content providers (if an ISP is vertically integrated) or, anyway, charge excessive fees to exploit CAPs. In this case, some have suggested to define data transmission and termination as relevant markets and cost-oriented prices as a remedy for SMP operators.[38] Indeed, a direct restraint

[36] Greenstein et al. (2017).

[37] Cave and Vogelsang (2015).

[38] Larouche (2012).

on ISPs' market power, taking into account the two-sides of the access market,[39] would represent a less distortive solution and would be more consistent with the overall regulatory framework than any net neutrality measure.[40]

Another issue is related to the imposition on end-users of data cap usage, which is not covered by net neutrality measures and is usually allowed because retail markets are considered competitive and thus are not regulated. Moreover, a data cap could well be a measure to avoid congestion, yet also create a (artificial) scarcity, making different contents work as substitutes for end-users and, thus, intensify the competition among content providers.[41]In this situation, data caps could reinforce the exploitative effect of a termination fee and prioritisation (absent net neutrality provisions) as CAPs would compete fiercely for the fast lane.

Another competitive concern is the discrimination within a *data cap strategy*. A vertically integrated ISP would exempt its own traffic, applying a *zero-rating* approach, in order to favour its content. A connected yet different issue is that of *zero-rating* agreements with a third-party CAP. The meaning of this agreement depends very much on the relative market power of the ISP and the CAP involved. It can be either (a) the result of an exploitation of market power, similar to the fast lane in a scarcity situation, by the ISP which would receive a payment for the agreement; or (b) an imposition of a CAP holding *must have* contents, whose unavailability may cause end-users to decide to switch ISP, where it faces strong competition.

Here, the question arises of whether a competitive IAS market could make net neutrality rules redundant in terms of preventing anti-competitive behaviours. Recent theories have generally supported the idea that lifting net neutrality rules on competing platforms is welfare-increasing.[42] However, this outcome is not due to competition reducing the incentives of ISPs to discriminate between content providers, as in the voice call termination (see Sect. 5.3), but to the intense competition among ISPs resulting in better prices for end-users and lower overall price distortions. Moreover, competition may influence the desirability of net neutrality

[39] ISPs could also be considered a two-sided platform connecting content providers with end-users.

[40] Net Neutrality as a *zero-pricing* rule can be considered a constraint on the business model of the two-sided platform (the ISP) forcing it to adopt a one-sided business model, i.e. charging only users and not CAPs.

[41] Economides and Hermalin (2015).

[42] Bourreau et al. (2015).

provision concerning investments in broadband capacity and content innovation: both of them would increase compared to a net neutrality regime, when competition is intense and a payment fast lane is allowed.[43]

Another important, but different, aspect of the debate regards the *internet traditions* of freedom, openness, and equality. In some way, the Internet was developed under a net neutrality regime, not falling under a regulation but resulting from market interaction. However, as already mentioned, this vision does not properly account for the transformation Internet has undergone, both in terms of structure—centralised versus decentralised—and data traffic—symmetric versus asymmetric.

Nevertheless, in the US and the EU the net neutrality debate has been developed also under the conceptual framework of a universal, equal and non-discriminatory access to digital services and content, increasingly seen as a fundamental end-user right. Where this is concerned, the Open Internet Access Regulation, which was adopted in 2015 by the European institutions, did not actually modify the regulatory access regime but the rules regarding universal services and user rights.[44] However, differently from the universal service obligation concerning electronic communication services (see Sect. 4.2), the universality and non-discriminatory provisions of digital services and content do not entail regulatory obligations for the providers of those services, but for the network intermediaries.

A more symmetric arrangement would be, instead, to extend the scope and rationale of the *geo-blocking* regulation, and its prohibition on digital service and content providers to discriminate against and deny access to end-users.[45]

The Open Internet Access Regulation grants end-users with enforceable rights to access and distribute information, content and services.[46] In doing so, this forces ISPs to "treat all traffic equally ... without discrimination, restriction or interference, and irrespective of the sender and receiver,

[43] Bourreau et al. (2015).

[44] Regulation (EU) 2015/2120 laying down measures concerning open internet access.

[45] Regulation (EU) 2018/302 on addressing unjustified geo-blocking and other forms of discrimination based on customers' nationality, place of residence or place of establishment within the internal market. The geo-blocking regulation was adopted to break unjustified geographical barriers to online sales in the EU Digital Single Market. Audio-visual media services giving access to copyright-protected content, as well as transactions with no relevant cross-border elements are excluded.

[46] Art 3(1) OIAR—Safeguarding of open internet access.

the content accessed or distributed, the applications or services used or provided, or the terminal equipment used."[47] However, "reasonable" day-to-day traffic management practices are allowed as long as they are (a) transparent, (b) non-discriminatory, (c) proportionate and (d) not based on any commercial considerations but on objectively different technical quality of service requirements for specific traffic categories. In any case, these measures may be maintained no longer than is necessary and cannot involve deep packet inspection. Some practices are clearly considered non-reasonable, e.g., "block, slow down, alter, restrict, interfere with, degrade or discriminate between specific content, applications or services".

There are three exceptions to the rule where compliance with legal obligations, network integrity and congestion management in exceptional and temporary situations are involved.

Regulation also allows for the provision of *specialised services*, deemed as those services "optimised for specific content, applications or services … where the optimisation is necessary in order to meet requirements of the content, applications or services for a specific level of quality"[48] or, in other words, those services that need to be carried out at a specific level of quality that cannot be assured by the standard *best effort* delivery.[49] The regulation defines specific safeguards to be respected for the provision of specialised services to ensure that the open Internet is not negatively affected. Specialised services (a) can be satisfied by the network capacity residual to any IAS provided; (b) are not usable or offered as a replacement for internet access services; and (c) are not to the detriment of the availability or general quality of the internet access for end-users.

Furthermore, the regulation defines transparency obligations for providers of internet access services additional to those existing for electronic communication service providers (Sect. 4.2).[50] In particular, contracts for internet access services must include easily accessible, accurate, meaningful

[47] Art 3(3) OIAR.

[48] Art 3(5) OIAR.

[49] BEREC Guidelines on the Implementation by National Regulators of European Net Neutrality Rules, BoR (16) 127. In its guidelines, BEREC also provides a few examples of specialised services, e.g., VoLTE (high-quality voice calling on mobile networks); linear (live) broadcasting IPTV services with specific quality requirements; real-time health services (e.g. remote surgery) and some new M2M communications services. In addition, it specifies that network-slicing in 5G networks may be used to deliver specialised services.

[50] Art. 4 OIAR.

and comparable information, covering (a) any traffic management measures used, and any impact on the end-user (e.g., quality of internet access, end-user privacy and personal data protection); (b) any data caps, speed and other quality of service parameters which may in practice impact internet access; (c) how any specialised services, to which the end-user subscribes, might in practice affect the same end-user's internet access services; (d) the download and upload speed of internet access services (with different metrics depending on fixed or mobile network); and (e) the remedies available to the consumer in case of any regular discrepancy between the actual performance of the internet access service and the contractually agreed on one.

As for enforcement, national regulators are empowered to closely monitor market developments and assess traffic management, commercial agreements and the compliance with transparency obligations in order to ensure the availability of non-discriminatory and transparent internet access at levels of quality that reflect advances in technology. For this purpose, national regulators may impose minimum quality of service requirements on internet access providers and other appropriate measures to ensure that all end-users enjoy an open internet access service. They must report annually on their findings to the Commission and the BEREC.[51] Moreover, according to the regulation provisions, in order to enhance a consistent application of the regulation, in 2016, BEREC drew up detailed guidelines, of which national regulators must take strictly into account.[52]

Finally, one of the main points debated concerns the *zero-rating* practices. The regulation explicitly established the freedom to conclude agreements between ISPs and end-users relating to commercial and technical conditions, as well as IAS aspects regarding price, data volumes or speed, and any commercial practices. Nevertheless, such agreements and commercial practices must not represent a limitation in the exercising of end-user rights and, consequently, circumvent provisions safeguarding open internet access.[53]

In particular, under the BEREC guidelines, some practices are clearly prohibited—for example, those where all applications are blocked or

[51] Under art. 5 OIAR, NRAs have published so far three set of yearly report, which have been sent to the Commission and BEREC.

[52] BEREC Guidelines on the Implementation by National Regulators of European Net Neutrality Rules, BoR (16) 127.

[53] Recital 7 and art 3(2) OIAR.

slowed down once the data cap is reached, except for the zero-rated application(s). Other practices need a specific assessment to be carried out by NRAs using the following certain criteria: (a) whether the practices circumvent the general aims of the regulation (to "safeguard equal and non-discriminatory treatment of traffic" and to "guarantee the continued functioning of the internet ecosystem as an engine of innovation"); (b) the market positions of the ISPs and CAPs involved; (c) any distorting effects on end-user choice, both for applications and CAPs; and d) the scale of the practice and the existence of alternatives.

8.4 From Net Neutrality, to Next Neutrality

The decoupling of network and services enables modularity of services, often provided in a bundle by the same player. Services are often characterised by enhanced substitutability or complementarity and linked to extensive network effects, which are important in shaping markets and market power. Here, each market transaction in the digital ecosystem entails an array of interdependent bilateral relationships, namely: (a) user to user (or to *business user*[54]); (b) user—ISP; (c) ISP—OTT; and (d) user—OTT.

As illustrated in the previous Sect. 8.3, different concepts regarding non-discrimination and transparency principles can be applied to each of the above relationships, reflecting a different side of net neutrality, some already covered by existing rules and some not (yet). Should regulation focus only on one specific side, and thus a specific strategy and contractual or market power relation, it would neglect existing interdependences and trade-offs or, worse, unintentionally generate other contractual or market power situations.

Let us look at some examples. Preventing price differentiations (non-paid prioritisation) in those relations between providers and OTTs is a measure that undoubtedly protects consumers, but it also ends up creating asymmetries in the relations between providers and OTTs concerning the appropriability of profits resulting from the user's connectivity.

[54] In the meaning of Regulation EU 2019/1150 on promoting fairness and transparency for business users of online intermediation services, the so-called Platform-to-Business (P2B) regulation, that is, any private individual acting in a commercial or professional capacity or any legal person offering goods or services to consumers through online intermediation services.

Agreements between providers and OTTs (zero rating) allow the provider to share some of the OTT profits and to benefit the user but can also generate exclusionary effects for other OTTs. A further complication of these scenarios arises from the fact that providers' behaviours are not independent from the degree of competition in the internet access markets and, vis à vis, the countervailing power of the OTTs. Further, discriminations between users put in place by ISPs are prohibited by the net neutrality rules, which could result in not guaranteeing a non-discriminatory access to digital services and content if digital service and content providers discriminate among users (limited for only specific aspects and services by the geo-blocking regulation).

In the OTT-user relationship, price and service differentiation is a distinctive trait, a feature of digital platform business models. Moreover, the implicit exchange of free services with data (*zero pricing*) can advantage users in the short term, but the opportunity costs are often not perceived by users, as well as the long-term impact on competition and on the consumers themselves. In this sense, net neutrality should also be addressed from the OTT-user side, covering, for example: (a) data exchange, to be explicit and transparent, by informing consumers about what data is required for each specific service and what is the "extra-profit" in terms of data for OTTs; (b) algorithm profiling and decision-making, allowing for an enhanced transparency concerning the governance of interactions with users and between users, tackling implicit forms of discrimination; (c) a fairer distribution and participation of users in profits generated by digital data transactions, enhancing digital consumer information and bargaining power, by (c_i) allowing appropriability and portability of own profiled data, and (c_{ii}) facilitating a data supply-side comparison and aggregation by third party digital tools (as happens in other sectors[55] and partially also with ICSs comparison tools in the EECC).

These data aggregator would be third party to platforms, i.e., vertically non-integrated and not active in other market sides. Those digital comparison tool could (i) calculate the value of data implicitly exchanged by users and (ii) monetise that data by either negotiating on behalf of users to obtain more favourable conditions with the platforms,

[55] For example, collective switching, coordinated by third parties, e.g., also digital comparison tool, in the energy sector in UK. See CMA (2016a) and CMA (2016b).

counterbalancing their bargaining power or by compering different data rewarding conditions in different platforms and guide the choice for a collective switching.

Some of these aspects have recently been addressed in different European legislations empowering business users and digital consumers,[56] whereas some other crucial aspects have still to be dealt with. As for business users, platform-to-business (P2B) regulation promotes fairness and transparency for business users of online intermediation services (between business users and consumers), by mandating (a) internal complaint-handling mechanisms to ensure ranking transparency and prohibit certain unfair practices, such as changes to online terms and conditions without a justifiable reason; (b) transparency and fairness obligations on contractual terms and conditions (e.g., notice period for term changes and termination);[57] (c) the ranking of transparency, making the up-to-date parameters on the basis of which rankings are formed visible to the public (while the actual functioning of the algorithm does not need to be disclosed in order to safeguard innovation incentives for platforms);[58] (d) the provision of an exhaustive justification for any heterogeneous form of differentiated treatment between certain business users and others on their platform.

As far as digital consumers are concerned, starting from a communication for a "new deal for consumers" in 2018,[59] the European Union has undertaken a modernisation of some aspects of the consumer protection rules to accommodate the digital ecosystem. involving some significant systemic net neutrality measures, despite formally being ascribed to the consumer and data protection domains.[60] These measures entail:

(a) ranking of consumer online search results, where digital service providers must inform consumers on the main parameters determining their ranking and relative importance and, moreover, any

[56] De Streel and Sibony (2017); Colangelo and Maggiolino (2019); Howells (2020).

[57] Art. 3 P2BR.

[58] Art. 5 P2BR.

[59] European Commission (2018) Communication: A New Deal for Consumers—COM/2018/0183 final.

[60] Directive (EU) 2019/2161 amending Directive 93/13/EEC (unfair contractual terms) and Directives 98/6/EC (consumer protection in the indication of the prices of products), 2005/29/EC (unfair commercial practices) and 2011/83/EU (consumer rights) as regards the better enforcement and modernisation of Union consumer protection rules (modernisation of digital consumers).

paid advertisement or payment received specifically for achieving a higher ranking within a specific search result must be clearly disclosed;[61]

(b) digital content and digital services with no monetary price but provided to digital consumers in exchange for personal data, included in the extended scope of the consumer rights directive;[62]

(c) personalised prices, of which consumers must be informed if those prices are personalised based on automated decision-making profiling their behaviour.[63]

Similarly, data portability rights, included within the General Data Protection Regulation (GDPR),[64] can be read as a necessary but insufficient complement to a policy aimed at balancing user-OTT interaction, by enabling switching of service providers. It encompasses three different and complementary rights: the right to receive data provided; the right to move that data to another data controller;[65] and the right to have the personal data transferred directly from one controller to another.[66]

A future-proof call for a *neutral network* should involve a systemic and holistic response to issues originating from digitalisation, multimedia convergence and the modular structure of digital ecosystem.

Here, the idea of *ne*[x]*t neutrality* rules should cover all possible contexts of opacity and non-discrimination in each of the relationships constituting the digital transaction, by assessing substitutability and complementarity of services and preventing or counteracting exertions of power across the whole digital ecosystem. Ne[x]t neutrality rules and implementation are therefore much closer to market analysis and

[61] Recitals 20—23 Directive (EU) 2019/2161.

[62] Recitals 31—33 Directive (EU) 2019/2161. As it is also the case for Directive (EU) 2019/770, which provides common rules on certain requirements concerning contracts for the supply of digital content or digital services, in particular on: (a) the conformity of digital content or a digital service with the contract, (b) remedies in the event of a lack of such conformity or a failure to supply, and the modalities for the exercise of those remedies, and (c) the modification of digital content or a digital service.

[63] Recital 45 Directive (EU) 2019/2161.

[64] Regulation (EU) 2016/6791 on the protection of natural persons with regard to the processing of personal data and on the free movement of such data, and repealing Directive 95/46/EC (General Data Protection Regulation).

[65] According to the GDPR, the data controller is the body determining the purposes and means of the processing of personal data.

[66] Art. 20 GDPR.

pro-competitive regulation, including the key role played in the market by empowered consumers (see Sect. 4.1). Focusing on the systemic nature of digital transactions leads to interpreting the net neutrality policy as a means to find a *balance of power* in the digital ecosystem, which will result in different regulatory measures, with different strength, depending on the actual market structure and dynamics.

This systemic policy approach cannot be effectively pursued through a fragmented and overly differentiated regulatory framework. Indeed, the European institutions adopted over time comprehensive policy frameworks for the digital ecosystem: the 2010 Digital Agenda for Europe and the 2015 Digital Single Market Strategy (see Box 1.1).

Nevertheless, the related EU legislation was mainly made up of a plurality of pieces of legislation, not fully coordinated. It is only in the last few years that a move towards systematisation and coordination has started and the recent 2020 Strategy, "shaping Europe's digital future,"[67] seems to be further extending the holistic approach towards the digital ecosystem, covering all aspects and legislative domains touched by the digital transformation. Under the 2020 Strategy, the EU institutions are (re)considering in a coordinated fashion: (a) a legislative framework for data governance and possibly a data act; (b) the EU competition law for the digital age; (c) an ex-ante, strengthened and modernised regulatory framework applicable to digital services (i.e., a *Digital Service Act* package); (d) a new Consumer Agenda empowering consumers to make informed choices and play an active role in the digital transformation.

This approach is in line with proposals advanced by many commentators to develop a coordinated regulation of digital services to ensure consistency, a level playing field and technological neutrality across the whole digital ecosystem. A single regulatory framework to be applied to "any service provided at a distance, using electronic communications networks", along with the already existing single framework for digital electronic communications networks.[68]

Here, a coordinated institutional framework also for all relevant national and supranational public bodies is necessary. Indeed, the intricate web of institutional interactions described regarding the telecom competition

[67] European Commission (2020).

[68] De Streel and Larouche (2016).

policy (Sect. 3.4) may turn into a tangled knot in the modular digital ecosystem. Digital transactions need a composite institutional order, governing and coordinating pro-competitive market regulation, personal data strategy and consumer empowerment. This objective can either be accomplished through an enhanced permanent coordination of all the public actors involved at national level,[69] or, as some proposals have advanced, by setting up *national authorities for digital services and networks*, internalising all externalities in the actions of different authorities, as well as minimising duplication costs, and increasing legal certainty.

In this regard, a 2019 joint survey by the Italian competition authority (AGCM), electronic communications regulator (AGCOM) and data protection authority (GPDP) stressed the importance of a "permanent coordination" between public regulators because an effective Big Data and digital economy policy should be grounded not only on law enforcement but also on coherent advocacy activities.

Two important reports in 2019, one in the UK from the Digital Competition Expert Panel and one in the US from the Stigler Centre at the University of Chicago, argue that a Digital Authority could be a valuable complement to antitrust enforcement. This new entity should establish a digital platform code of conduct, oversee personal data mobility and systems with open standards, addressing bottleneck power, and create "light touch" rules (behavioural nudges) that can lead consumers to make better choices and pursue data openness as a tool to increase competition.[70]

A Digital Authority would embrace all the different interrelated aspects and actors of digital transactions, focusing on power bottlenecks, users and data mobility, and the interoperability of all kinds of networks characterised by pervasive network effects, thus, pursuing and fully enabling a *ne*[x]*t neutrality* future-proof approach for the digital ecosystem.

The debate, here, is just at the beginning.

[69] AGCM, AGCOM, GPDP (2019).

[70] The Stigler Committee for the Study of Digital Platforms (2019). British Digital Competition Expert (2019).

References

AGCM, AGCOM, GPDP. (2019). Joint survey on Big Data.

AGCOM. (2018). Interim Report in the Context of the Joint Inquiry on Big Data.

AGCOM. (2019). Report on Online Platforms.

Armstrong, M. (2006). Competition in Two-Sided Markets. *The Rand Journal of Economics, 37*(3), 668–691.

BEREC. (2013). Report on the NRAs' Regulatory Capacity.

BEREC. (2016). Report on OTT Services—BoR (16) 35.

BEREC. (2018). European Benchmark of the pricing of bundles, BoR (18)171.

Bourreau, M., Kourandi, F., & Valletti, T. (2015). Net Neutrality with Competing Internet Platforms. *Journal of Industrial Economics, 63*(1), 30–73.

British Digital Competition Expert. (2019). *Unlocking Digital Competition* (Furman Report).

Cave, M., & Vogelsang, I. (2015). Net Neutrality: An E.U./U.S. Comparison. *Competition Policy International, 11*(1), 85–95.

CMA. (2016a). Digital Comparison Tool Market Study Digital Comparison Tool Market Study.

CMA. (2016b). Energy Market Investigation.

Colangelo, G., & Maggiolino, M. T. (2019). From Fragile to Smart Consumers: Shifting Paradigm for the Digital Era. *Computer Law and Security Review, 35*(2), 173–181.

De Streel, A., & Larouche, P. (2016). *An Integrated Regulatory Framework for Digital Network and Services.* CERRE Policy Report.

De Streel, A., & Sibony, A. (2017). *Toward Smarter Consumer Protection Rules for Digital Society.* CERRE Report.

Delmastro, M., & Nicita, A. (2019). *Big Data.* Il Mulino Publishing.

Economides, N., & Hermalin, E. (2015). The Strategic Use of Download Limits by a Monopoly Platform. *RAND Journal of Economics, 46*(2), 297–327.

European Commission. (1997). Green Paper on the Convergence of Telecommunications, Media and Information Technology Sectors, and Implications for Regulation, COM(97)623.

European Commission. (2016). Communication on Online Platforms and the Digital Single Market Opportunities and Challenges for Europe—COM/2016/0288 Final.

European Commission. (2018). Digital Scoreboard.

European Commission. (2019). Digital Agenda Scoreboard.

European Commission. (2020). Communication: A European Strategy for Data—COM(2020) 66 Final.

Filistrucchi, L., Geradin, D., & van Damme, E. (2013). Identifying Two-Sided Markets. *World Competition, 36*(1), 33–59.

Greenstein, S., Peitz, M., & Valletti, T. (2017). Net Neutrality: A Fast Lane to Understanding the Trade-Offs. *Journal of Economic Perspectives, 30*(2), 127–150.
Howells, G. (2020). Protecting Consumer Protection Values in the Fourth Industrial Revolution. *Journal of Consumer Policy, 43*, 145–175.
Larouche, P. (2012). *Network Neutrality: The Global Dimension* (Eds. M. Burri & T. Cottier). Trade Governance in the Digital Age: World Trade Forum, pp. 91–122.
OECD. (2009). Better Regulation in Europe: An Assessment of Regulatory Capacity in 15 Member States of the European Union.
OECD. (2014). *Connected Televisions: Convergence and Emerging Business Models.* OECD Digital Economy Papers, No. 231.
OECD. (2015). *Triple and Quadruple Play Bundles of Communication Services.* OECD Science, Technology and Industry Policy Papers, No. 23.
OECD. (2016). Digital Convergence and Beyond, Digital Economy Papers No. 251
Rochet, J. C., & Tirole, J. (2006). Two-Sided Markets: A Progress Report. *The Rand Journal of Economics, 37*(3), 645–667.
Schiller, D. (2000). *Digital Capitalism.* Networking the Global Market System.
Stucke, M. (2016). *Big Data and Competition Policy.* Oxford University Press.
The Stigler Committee for the Study of Digital Platforms. (2019). Market Structure and Antitrust Subcommittee.
Wu, T. (2003). Network Neutrality, Broadband Discrimination. *Journal of Telecommunications and High Technology Law, 2*(1), 141–178.

The manufacturer's authorised representative in the EU is Springer Nature Customer Service Centre GmbH, Europaplatz 3, 69115 Heidelberg, Germany. If you have any concerns regarding our products, please contact ProductSafety@springernature.com

Printed and bound by CPI Group (UK) Ltd, Croydon, CR0 4YY
30/09/2026
02225461-0001